Displaying Time Series, Spatial, and Space-Time Data with R

Focusing on the exploration of data with visual methods, "Displaying Time Series, Spatial, and Space-Time Data with R" presents methods and R code for producing high-quality static graphics, interactive visualizations, and animations, of time series, spatial, and space-time data. Practical examples using real-world datasets help you understand how to apply the methods and code.

While the first edition primarily focused on static graphics, and the second edition significantly expanded upon interactive elements, this edition has been thoroughly updated with a modern emphasis. Key enhancements include:

- Comprehensive updates to all chapters: Incorporating the latest advancements in packages like "sf" and "terra."
- New content: Introducing valuable topics such as trajectories, hill shading, and cartograms, enriching the scope of the book.

The book illustrates how to display a dataset starting with an easy and direct approach and progressively adding improvements that involve more complexity. Each of the three parts of the book is devoted to different types of data. In each part, the chapters are grouped according to the various visualization methods or data characteristics.

Features

- Offers detailed information on producing high-quality graphics, interactive visualizations, and animations
- Uses real data from meteorological, climate, economic, social science, energy, engineering, environmental, and epidemiological research in many practical examples
- Shows how to improve graphics based on visualization theory
- Provides the graphics, data, and R code on the author's website, enabling you to practice with the methods and modify the code to suit your own needs.

Chapman & Hall/CRC
The R Series

Series Editors
John M. Chambers, Department of Statistics, Stanford University, California, USA
Torsten Hothorn, Division of Biostatistics, University of Zurich, Switzerland
Duncan Temple Lang, Department of Statistics, University of California, Davis, USA
Hadley Wickham, RStudio, Boston, Massachusetts, USA

Recently Published Titles

Deep Learning and Scientific Computing with R torch
Sigrid Keydana

Model-Based Clustering, Classification, and Density Estimation Using mclust in R
Lucca Scrucca, Chris Fraley, T. Brendan Murphy, and Adrian E. Raftery

Spatial Data Science: With Applications in R
Edzer Pebesma and Roger Bivand

Modern Data Visualization with R
Robert Kabacoff

Learn R: As a Language, Second Edition
Pedro J. Aphalo

Spatial Analysis in Geology Using R
Pedro M. Nogueira

Analyzing Baseball Data with R, Third Edition
Jim Albert, Benjamin S. Baumer and Max Marchi

Geocomputation with R, Second Edition
Robin Lovelace, Jakub Nowosad, Jannes Muenchow

Microeconometrics with R
Yves Croissant

Statistical Inference via Data Science
A ModernDive into R and the Tidyverse, Second Edition
Chester Ismay, Albert Y. Kim, and Arturo Valdivia

Copula Additive Distributional Regression Using R
Giampiero Marra and Rosalba Radice

Introduction to Political Analysis in R
H. Whitt Kilburn

Displaying Time Series, Spatial, and Space-Time Data with R
Oscar Perpiñán Lamigueiro

For more information about this series, please visit: https://www.crcpress.com/Chapman--HallCRC-The-R-Series/book-series/CRCTHERSER

Displaying Time Series, Spatial, and Space-Time Data with R

Oscar Perpiñán Lamigueiro

CRC Press
Taylor & Francis Group
Boca Raton London New York

CRC Press is an imprint of the
Taylor & Francis Group, an **informa** business

A CHAPMAN & HALL BOOK

Designed cover image: Oscar Perpiñán Lamigueiro

Third edition published 2025
by CRC Press
2385 NW Executive Center Drive, Suite 320, Boca Raton FL 33431

and by CRC Press
4 Park Square, Milton Park, Abingdon, Oxon, OX14 4RN

CRC Press is an imprint of Taylor & Francis Group, LLC

Library of Congress Cataloging-in-Publication Data
Names: Perpiñán Lamigueiro, Óscar, author.
Title: Displaying time series, spatial, and space-time data with R / Oscar Perpiñán Lamigueiro.
Description: Third edition. | Boca Raton : CRC Press, 2025. | Includes bibliographical references and index. | Summary: "Focusing on the exploration of data with visual methods, "Displaying Time Series, Spatial, and Space-Time Data with R" presents methods and R code for producing high-quality static graphics, interactive visualizations, and animations, of time series, spatial, and space-time data. Practical examples using real-world datasets help you understand how to apply the methods and code. While the first edition primarily focused on static graphics, and the second edition significantly expanded upon interactive elements, this edition has been thoroughly updated with a modern emphasis. Key enhancements include: (1) Comprehensive updates to all chapters: Incorporating the latest advancements in packages like "sf" and "terra." (2) New content: Introducing valuable topics such as trajectories, hill shading, and cartograms, enriching the scope of the book. The book illustrates how to display a dataset starting with an easy and direct approach and progressively adding improvements that involve more complexity. Each of the three parts of the book is devoted to different types of data. In each part, the chapters are grouped according to the various visualization methods or data characteristics"-- Provided by publisher.
Identifiers: LCCN 2025014326 (print) | LCCN 2025014327 (ebook) | ISBN 9781032776217 (hbk) | ISBN 9781032779201 (pbk) | ISBN 9781003485384 (ebk)
Subjects: LCSH: Time-series analysis--Data processing. | R (Computer program language)
Classification: LCC QA280 .P475 2025 (print) | LCC QA280 (ebook) | DDC 519.5/502855133--dc23/eng/20250415
LC record available at https://lccn.loc.gov/2025014326
LC ebook record available at https://lccn.loc.gov/2025014327

ISBN: 978-1-032-77621-7 (hbk)
ISBN: 978-1-032-77920-1 (pbk)
ISBN: 978-1-003-48538-4 (ebk)

DOI: 10.1201/9781003485384

Typeset in URW PalladioL font
by KnowledgeWorks Global Ltd.

Contents

List of Figures

Chapter 1

Introduction

1.1 What This Book Is About

A data graphic is not only a static image but also tells a story about the data. It activates cognitive processes that are able to detect patterns and discover information not readily available with the raw data. This is particularly true for time series, spatial, and space-time datasets.

There are several excellent books about data graphics and visual perception theory, with guidelines and advice for displaying information, including visual examples. Let's mention *The Elements of Graphical Data* (Cleveland 1994) and *Visualizing Data* (Cleveland 1993) by W. S. Cleveland, *Envisioning Information* (Tufte 1990) and *The Visual Display of Quantitative Information* (Tufte 2001) by E. Tufte, *The Functional Art* by A. Cairo (Cairo 2012), and *Visual Thinking for Design* by C. Ware (Ware 2021). Ordinarily, they do not include the code or software tools to produce those graphics.

On the other hand, there is a collection of books that provides code and detailed information about the graphical tools available with R. Commonly they do not use real data in the examples and do not provide advice for improving graphics according to visualization theory. Three books are the unquestioned representatives of this group: *R Graphics* by P. Murrell (Murrell 2021), *Lattice: Multivariate Data Visualization with R* by D. Sarkar (Sarkar 2008), and *ggplot2: Elegant Graphics for Data Analysis* by H. Wickham (Wickham 2016).

DOI: 10.1201/9781003485384-1

This book proposes methods to display time series, spatial, and space-time data using R, and aims to be a synthesis of both groups, providing code and detailed information to produce high-quality graphics with practical examples.

1.2 What You Will *Not* Find in This Book

- **This is not a book to learn R.**

  Readers should have a fair knowledge of programming with R to understand the book. In addition, previous experience with the zoo, sp, raster, lattice, ggplot2, and grid packages is helpful.

  If you need to improve your R skills, consider these information sources:

  - Introduction to R[1].
  - Official manuals[2].
  - Contributed documents[3].
  - Mailing lists[4].
  - R-bloggers[5].
  - Books related to R[6] and particularly *Software for Data Analysis* by John M. Chambers (Chambers 2008).

- **This book does not provide an exhaustive collection of visualization methods.**

  Instead, it illustrates what I found to be the most useful and effective methods. Notwithstanding, each part includes a section titled "Further Reading" with bibliographic proposals for additional information.

- **This book does not include a complete review or discussion of R packages.**

---

[1] http://cran.r-project.org/doc/manuals/R-intro.html
[2] http://cran.r-project.org/manuals.html
[3] http://cran.r-project.org/other-docs.html
[4] http://www.r-project.org/mail.html
[5] http://www.r-bloggers.com
[6] http://www.r-project.org/doc/bib/R-books.html

Their most useful functions, classes, and methods regarding data and graphics are outlined in the introductory chapter of each part and conveniently illustrated with the help of examples. However, if you need detailed information about a certain aspect of a package, you should read the corresponding package manual or vignette. Moreover, if you want to know additional alternatives, you can navigate through the CRAN Task Views about Time Series[7], Spatial Data[8], Spatiotemporal Data[9], and Dynamic Graphics[10].

- **Finally, this book is not a handbook of data analysis, geostatistics, point pattern analysis, or time series theory**.

  Instead, this book is focused on the exploration of data with visual methods, so it may be framed in the Exploratory Data Analysis approach. Therefore, this book may be a useful complement to superb bibliographic references where you will find plenty of information about those subjects. For example, (Chatfield 2019), (Cressie and C. Wikle 2015), (Slocum 2022) and (Bivand, E. J. Pebesma, and Gomez-Rubio 2013).

1.3 How to Read This Book

This book is organized into three parts, each devoted to different types of data. Each part comprises several chapters according to the various visualization methods or data characteristics. The chapters are structured as independent units so readers can jump directly to a certain chapter according to their needs. Of course, there are several dependencies and redundancies between the sets of chapters that have been conveniently signaled with cross-references.

The content of each chapter illustrates how to display a dataset starting with an easy and direct approach. Often, this first result is not entirely satisfactory, so additional improvements are progressively added. Each step involves additional complexity that in some cases, can be overwhelming during a first reading. Thus, some sections, marked with the sign ☜, can be safely skipped for later reading.

Although I have done my best to help readers understand the methods and code, you should not expect to understand it after one reading. The

---

[7]http://cran.r-project.org/view=TimeSeries
[8]http://cran.r-project.org/view=Spatial
[9]http://cran.r-project.org/view=SpatioTemporal
[10]https://cran.r-project.org/view=DynamicVisualizations

key is practical experience, and the best way is to try out the code with the provided data **and** modify it to suit your needs with your own data. There is a website and a code repository to help you in this task.

1.3.1 Website and Code Repository

The book website with the main graphics of this book is located at

https://oscarperpinan.codeberg.page/bookvis

The full code is freely available from the repository:

https://codeberg.org/oscarperpinan/bookvis

On the other hand, the datasets used in the examples are either available at the repository or can be freely obtained from other websites. It must be underlined that the combination of code and data freely available allows this book to be fully reproducible.

I have chosen the datasets according to two main criteria:

- They are freely available without restrictions for public use.

- They cover different scientific and professional fields (meteorology and climate research, economy and social sciences, energy and engineering, environmental research, epidemiology, etc.).

The repository and the website can be downloaded as a compressed file[11], or if you use git, you can clone the repository with:

```
git clone https://codeberg.org/oscarperpinan/bookvis.git
```

1.4 R Graphics

There are two distinct graphics systems built into R, referred to as traditional and grid graphics. Grid graphics are produced with the grid package (Murrell 2021), a flexible low-level graphics toolbox. Compared with the traditional graphics model, it provides more flexibility to modify or add content to an existent graphical output, better support for combining different outputs easily, and more possibilities for interaction. All the graphics in this book have been produced with the grid graphics model.

Other packages are constructed over it to provide high-level functions, most notably the lattice and ggplot2 packages.

---

[11]https://codeberg.org/oscarperpinan/bookvis/archive/master.zip

1.4.1 lattice

The `lattice` package (Sarkar 2008) is an independent implementation of Trellis graphics, which were mostly influenced by *The Elements of Graphing Data* (Cleveland 1994). Trellis graphics often consist of a rectangular array of panels. The `lattice` package uses a *formula* interface to define the structure of the array of panels with the specification of the variables involved in the plot. The result of a `lattice` high-level function is a `trellis` object.

For bivariate graphics, the formula is generally of the form y ~ x representing a single panel plot with y versus x. This formula can also involve expressions. The main function for bivariate graphics is `xyplot`.

Optionally, the formula may be y ~ x | g1 * g2 and y is represented against x conditional on the variables g1 and g2. Each unique combination of the levels of these conditioning variables determines a subset of the variables x and y. Each subset provides the data for a single panel in the Trellis display, an array of panels laid out in columns, rows, and pages.

For example, in the following code, the variable `wt` of the dataset `mtcars` is represented against the `mpg`, with a panel for each level of the categorical variable `am`. The points are grouped by the values of the `cyl` variable.

```
xyplot(wt ~ mpg | am, data = mtcars, groups = cyl)
```

For trivariate graphics, the formula is of the form z ~ x * y, where z is a numeric response, and x and y are numeric values evaluated on a rectangular grid. Once again, the formula may include conditioning variables, for example, z ~ x * y | g1 * g2. The main function for these graphics is `levelplot`.

The plotting of each panel is performed by the panel function, specified in a high-level function call as the `panel` argument. Each high-level `lattice` function has a default panel function, although the user can create new Trellis displays with custom panel functions.

`lattice` is a member of the recommended packages list, so it is commonly distributed with R itself. There are more than 250 packages depending on it, and the most important packages for our purposes (`zoo`, `sp`, and `raster`) define methods to display their classes using `lattice`.

On the other hand, the `latticeExtra` package (Sarkar and Andrews 2022) provides additional flexibility for the somewhat rigid structure of the Trellis framework implemented in `lattice`. This package complements the `lattice` with the implementation of layers via the `layer` function, and superposition of `trellis` objects and layers with the `+.trellis` function.

Using both packages, you can define a graphic with the formula interface (under the `lattice` model) and overlay additional content as layers (following the `ggplot2` model).

1.4.2 ggplot2

The `ggplot2` package (Wickham 2016) is an implementation of the system proposed in *The Grammar of Graphics* (Wilkinson 2005), a general scheme for data visualization that breaks up graphs into semantic components such as scales and layers. Under this framework, the definition of the graphic with `ggplot2` is done with a combination of several functions that provides the components, instead of the formula interface of `lattice`.

With `ggplot2`, a graphic is composed of:

- A dataset, `data`, and a set of mappings from variables to aesthetics, `aes`.

- One or more layers, each composed of: a geometric object, `geom_*`, to control the type of plot you create (points, lines, etc.); a statistical transformation, `stat_*`; and a position adjustment (and optionally, additional dataset and aesthetic mappings).

- A scale, `scale_*`, to control the mapping from data to aesthetic attributes. Scales are common across layers to ensure a consistent mapping from data to aesthetics.

- A coordinate system, `coords_*`.

- Optionally, a faceting specification, `facet_*`, the equivalent of Trellis graphics with panels.

The function `ggplot` is typically used to construct a plot incrementally, using the + operator to add layers to the existing ggplot object. For instance, the following code (equivalent to the previous `lattice` example) uses `mtcars` as the dataset and maps the `mpg` variable on the x-axis and the `wt` variable on the y-axis. The geometric object is the point using the `cyl` variable to control the color. Finally, the levels of the `am` variable define the panels of the graphic.

```
ggplot(mtcars, aes(mpg, wt)) +
    geom_point(aes(colour=factor(cyl))) +
    facet_grid(. ~ am)
```

This package is very popular, with a large list of packages depending on it. In the context of this book, time series can be displayed with it because the zoo package defines the autoplot function based on ggplot2. Regarding spatial data, this package provides a geom function designed for spatial data. Detailed information is provided in Section 7.1.2.

1.4.3 Comparison between lattice and ggplot2

Which package to choose is, for a wide range of datasets, a question of personal preferences. You may be interested in a comparison between them published in a series of blog posts[12]. Consequently, where possible, most of the code contains alternatives defined both with lattice and with ggplot2.

It is important to note that both latticeExtra and ggplot2 define a function named layer. The ggplot2::layer function is rarely called by the user, because the wrapper functions geom_* and stats_ are preferred. On the other hand, the latticeExtra::layer function is designed to be directly called by the user, and therefore its masking must be prevented. Consequently, when the latticeExtra and ggplot2 packages are to be working together in the same session, the latticeExtra package must be loaded after ggplot2.

1.4.4 Interactive graphics

Both lattice and ggplot2 (and every package based on grid) generate static graphics. However, interactive web graphics produced with R have experienced a boost in recent years, mainly thanks to the package htmlwidgets (Vaidyanathan et al. 2023). This package provides a framework for creating R bindings to JavaScript libraries. This package is the base for important visualization packages such as dygraphs, highcharter, plotly, leaflet, and mapview. They will be covered along the chapters of the book.

On the other hand, the package gridSVG (Murrell and Potter 2023) converts any grid scene to a Scalable Vector Graphics (SVG) document. The grid.hyperlink function allows a hyperlink to be associated with any component of the scene, the grid.animate function can be used to animate any component of a scene and the grid.garnish function can be used to add SVG attributes to the components of a scene. By setting event

---

[12]http://learnr.wordpress.com/2009/06/28/

handler attributes on a component, plus possibly using the grid.script function to add JavaScript to the scene, it is possible to make the component respond to user input such as mouse clicks.

1.5   Packages

Throughout the book, several R packages are used. All of them are available from CRAN, and you must install them before using the code. Most of them are loaded at the start of the code of each chapter, although some of them are loaded later if they are used only inside optional sections (marked with 🜨). You should install the last version available at CRAN to ensure correct functioning of the code.

Although the introductory chapter of each part includes a section with an outline of the most relevant packages, some of them deserve to be highlighted here:

- zoo (Zeileis and Grothendieck 2005) provides infrastructure for time series using arbitrary classes for the time stamps (Section 2.1.1).

- sp (E. Pebesma 2012) and sf (E. Pebesma 2024) provide a coherent set of classes and methods for the major spatial data types: points, lines, polygons, and grids (Sections 7.1.1 and 7.1.2). spacetime (E. Pebesma 2012) and sftime (Teickner, Edzer Pebesma, and Graeler 2024) define classes and methods for spatiotemporal data, and methods for plotting data as map sequences or multiple time series (Sections 16.1.1 and 16.1.2).

- raster (R. J. Hijmans 2024a) and its successor terra (R. J. Hijmans 2024b) are a major extension of gridded spatial data classes. They provide a unified access method to different raster formats, permitting large objects to be analyzed with the definition of basic and high-level processing functions (Sections 7.1.3 and 16.1.3). raster-Vis (O. Perpiñán and R. Hijmans 2023) provides enhanced visualization of raster data with methods for spatiotemporal rasters (Sections 7.1.5 and 16.1.5).

1.6   Software Used to Write This Book

This book has been written using different computers running Debian GNU Linux and using several gems of open-source software:

- org-mode (Schulte et al. 2012), LaTeX, and AUCTeX, for authoring text and code.

- R (R Core Team 2024) with Emacs Speaks Statistics (Rossini et al. 2004).

- GNU Emacs as development environment[13].

1.7 About the Author

Throughout my professional career, my main area of expertise has been photovoltaic solar energy systems, with a special interest in solar radiation. Initially I worked as an engineer for a private company, and I was involved in several commercial and research projects. The project teams were partly integrated by people with low technical skills who relied on the input from engineers to complete their work. I learned how a good visualization output eased the communication process.

Now I work as a professor and researcher at the university. Data visualization is one of the most important tools I have available. It helps me embrace and share the steps, methods, and results of my research. With students, it is an inestimable partner in helping them understand complex concepts.

I have been using R to simulate the performance of photovoltaic energy systems and to analyze solar radiation data, both as time series and spatial data. As a result, I have developed packages that include several graphical methods to deal with multivariate time series (namely, solaR (O. Perpiñán 2012), meteoForecast (Oscar Perpiñán and Almeida 2023), and PVF (Pinho-Almeida, O. Perpiñán, and Narvarte 2015)) and space-time data (rasterVis (O. Perpiñán and R. Hijmans 2023)).

1.8 Acknowledgments

Writing a book is often described as a solitary activity, but while writing or coding, I feel immersed in a huge collaborative network of past and present contributors. Piotr Kropotkin described it with the following words (Kropotkin 1906):

> Thousands of writers, poets, scholars, have laboured to increase knowledge, dissipate error, and create that atmosphere

---

[13]https://www.gnu.org/savannah-checkouts/gnu/emacs/emacs.html

of scientific thought, without which the marvels of our century could never have appeared. And these thousands of philosophers, poets, scholars, inventors, have themselves been supported by the labour of past centuries. They have been upheld and nourished through life, both physically and mentally, by legions of workers and craftsmen of all sorts.

And Lewis Mumford claimed (Mumford 1934):

Socialize Creation! What we need is the realization that the creative life, in all its manifestations, is necessarily a social product.

I want to express my deepest gratitude and respect to all those women and men who have contributed and contribute to strengthening the communities of free software, open data, and open science. My special thanks go to the people of the R community: users, members of the R Core Development Team, and package developers.

With regard to this book in particular, I would like to thank John Kimmel and Lara Spieker for the constant support, guidance, and patience.

Last, and most importantly, thanks to Candela, Marina, and Javi, my permanent source of happiness, imagination, and love. Thanks to María, *mi amor, mi cómplice y todo.*

Part I

Time Series

Chapter 2

Displaying Time Series: Introduction

A time series is a sequence of observations registered at consecutive time instants. When these time instants are evenly spaced, the distance between them is called the sampling interval. The visualization of time series is intended to reveal changes of one or more quantitative variables through time and to display the relationships between the variables and their evolution through time.

The standard time series graph displays the time along the horizontal axis. Several variants of this approach can be found in Chapter 3. On the other hand, time can be conceived as a grouping or conditioning variable (Chapter 4). This solution allows several variables to be displayed together with a scatterplot, using different panels for subsets of the data (time as a conditioning variable) or using different attributes for groups of the data (time as a grouping variable). Moreover, time can be used as a complementary variable that adds information to a graph where several variables are confronted (Chapter 5).

These chapters provide a variety of examples to illustrate a set of useful techniques. These examples make use of several datasets (available at the book website) described in Chapter 6.

DOI: 10.1201/9781003485384-2

2.1 Packages

The CRAN Tasks View "Time Series Analysis"[1] summarizes the packages for reading, visualizing, and analyzing time series. This section provides a brief introduction to the zoo and xts packages. Most of the information has been extracted from their vignettes, webpages, and help pages. You should read them for detailed information.

Both packages extensively use the time classes defined in R. The interested reader will find an overview of the different time classes in R in (Ripley and Hornik 2001) and (Grothendieck and Petzoldt 2004).

2.1.1 zoo

The zoo package (Zeileis and Grothendieck 2005) provides an S3 class with methods for totally ordered indexed observations. Its key design goals are independence of a particular index class and consistency with base R and the ts class for regular time series.

Objects of class zoo are created by the function zoo from a numeric vector, matrix, or a factor that is totally ordered by some index vector. This index is usually a measure of time, but every other numeric, character, or even more abstract vector that provides a total ordering of the observations is also suitable. It must be noted that this package defines two new index classes, yearmon and yearqtr, for representing monthly and quarterly data, respectively.

The package defines several methods associated with standard generic functions such as print, summary, str, head, tail, and [(subsetting). In addition, standard mathematical operations can be performed with zoo objects, although only for the intersection of the indexes of the objects.

On the other hand, the data stored in zoo objects can be extracted with coredata, which drops the index information and can be replaced by coredata<-. The index can be extracted with index or time and can be modified by index<-. Finally, the window and window<- methods extract or replace time windows of zoo objects.

Two zoo objects can be merged by common indexes with merge and cbind. The merge method combines the columns of several objects along the union or the intersection of the indexes. The rbind method combines the indexes (rows) of the objects.

---

[1] http://CRAN.R-project.org/view=TimeSeries

The `aggregate` method splits a `zoo` object into subsets along a coarser index grid, computes a function (`sum` is the default) for each subset, and returns the aggregated `zoo` object.

This package provides four methods for dealing with missing observations:

1. `na.omit` removes incomplete observations.

2. `na.contiguous` extracts the longest consecutive stretch of non-missing values.

3. `na.approx` replaces missing values by linear interpolation.

4. `na.locf` replaces missing observations by the most recent non-NA prior to it.

The package defines interfaces to `read.table` and `write.table` for reading, `read.zoo`, and writing, `write.zoo`, zoo series from or to text files. The `read.zoo` function expects either a text file or connection as input or a `data.frame`. `write.zoo` first coerces its argument to a `data.frame`, adds a column with the index, and then calls `write.table`.

2.1.2 xts

The `xts` package (Ryan and Ulrich 2024) extends the `zoo` class definition to provide a general time-series object. The index of an `xts` object must be of a time or date class: `Date`, `POSIXct`, `chron`, `yearmon`, `yearqtr`, or `timeDate`. With this restriction, the subset operator `[` is able to extract data using the ISO:8601[2] time format notation `CCYY-MM-DD HH:MM:SS`. It is also possible to extract a range of times with a `from`/`to` notation, where both from and to are optional. If either side is missing, it is interpreted as a request to retrieve data from the beginning or through the end of the data object.

Furthermore, this package provides several time-based tools:

- `endpoints` identifies the endpoints with respect to time.

- `to.period` changes the periodicity to a coarser time index.

- The functions `period.*` and `apply.*` evaluate a function over a set of non-overlapping time periods.

---

[2]`http://en.wikipedia.org/wiki/ISO_8601`

2.2 Further Reading

- (Wills 2012) provides a systematic analysis of the visualization of time series, and a section of (Heer, Bostock, and Ogievetsky 2010) summarizes the main techniques to display time series.

- (Cleveland 1994) includes a section about time series visualization with a detailed discussion of the banking to 45° technique and the cut-and-stack method. (Heer and Agrawala 2006) propose multiscale banking, a technique to identify trends at various frequency scales.

- (Few 2008; Heer, Kong, and Agrawala 2009) explain in detail the foundations of the horizon graph (Section 3).

- The *small multiples* concept (Sections 3.2 and 4.1) is illustrated in (Tufte 2001; Tufte 1990).

- Stacked graphs are analyzed in (Byron and Wattenberg 2008), and the ThemeRiver technique is explained in (Havre et al. 2002).

- (Cleveland 1994; Friendly and Denis 2005) study the scatterplot matrices (Section 4.1), and (D. B. Carr et al. 1987) provide information about hexagonal binning.

- (Harrower and Fabrikant 2008) discuss the use of animation for the visualization of data. (Few 2007) exposes a software tool resembling the Trendalyzer.

- The D3 gallery[3] shows several great examples of time-series visualizations using the JavaScript library D3.js.

---

[3]https://observablehq.com/@d3/gallery

Chapter 3

Time on the Horizontal Axis

The most frequent visualization method of a time series uses the horizontal axis to depict the time index. This chapter illustrates several variants of this approach to display multivariate time series: multiple time series with different scales, variables with the same scale, and stacked graphs. The last section provides examples and code for producing interactive visualizations.

Along with this chapter, these subjects are covered: small multiples, panel functions, aspect ratio, diverging palettes, horizon and stacked graphs, and interactive visualization.

The most relevant packages in this chapter are: zoo and xts for reading and arranging data as time series; colorspace for defining color palettes; and dygraphs, highcharter, plotly, and streamgraph for interactive visualization.

DOI: 10.1201/9781003485384-3

3.1 Time Graph of Variables with Different Scales

There is a variety of scientific research interested in the relationship among several meteorological variables. A suitable approach is to display the time evolution of all of them using a panel for each of the variables. The superposition of variables with different characteristics is not very useful (unless their values were previously rescaled), so this option is postponed for Section 3.2.

For this example, we will use the eight years of daily data from the SIAR meteorological station located at Aranjuez (Madrid). This multivariate time series can be displayed with the xyplot method of lattice for zoo objects with a panel for each variable (Figure 3.1).

```
library("zoo")
load('data/TimeSeries/aranjuez.RData')

## The layout argument arranges panels in rows
xyplot(aranjuez, layout = c(1, ncol(aranjuez)))
```

The package ggplot2 provides the generic method autoplot to automate the display of certain classes with a simple command. The package zoo provides an autoplot method for the zoo class with a result similar to that obtained with xyplot (Figure 3.2)

```
autoplot(aranjuez) + facet_free()
```

3.1.1 ✎Annotations to Enhance the Time Graph

These first attempts can be improved with a custom panel function that generates the content of each panel using the information processed by xyplot or overlaying additional layers with autoplot. One of the main enhancements is to highlight certain time regions that fulfill certain conditions. The package latticeExtra provides a nice solution for xyplot with panel.xblocks. The result is displayed in Figure 3.3:

- The alternating of years is displayed with blocks of gray and white color using the panel.xblocks function from latticeExtra (line 11). The year is extracted (as a character) from the time index of the zoo object with format.POSIX1t (line 3).

- Those values below the mean of each variable are highlighted with short red color blocks at the bottom of each panel, again with the panel.xblocks function (line 15).

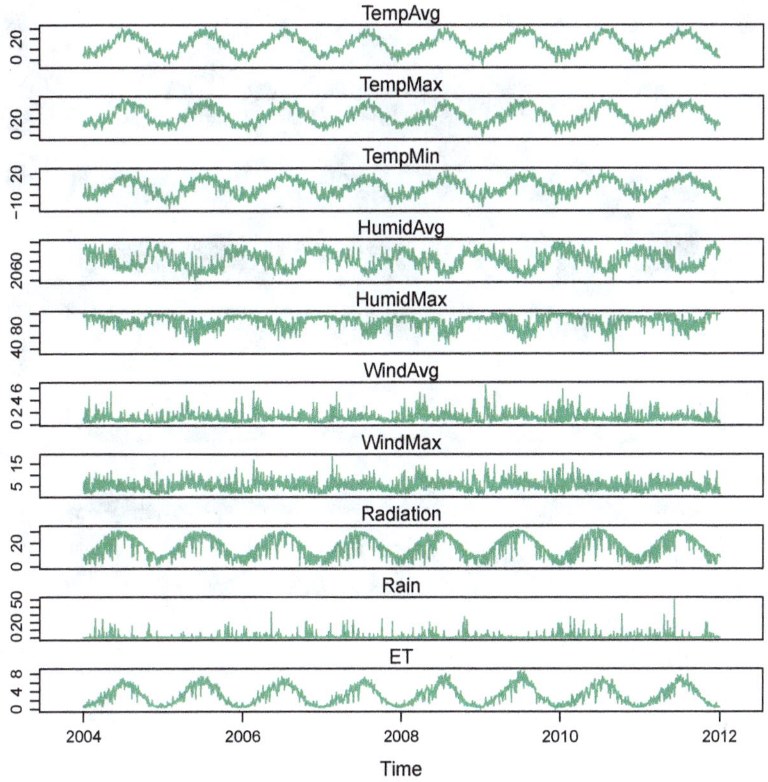

FIGURE 3.1: Time plot of the collection of meteorological time series of the Aranjuez station (`lattice` version).

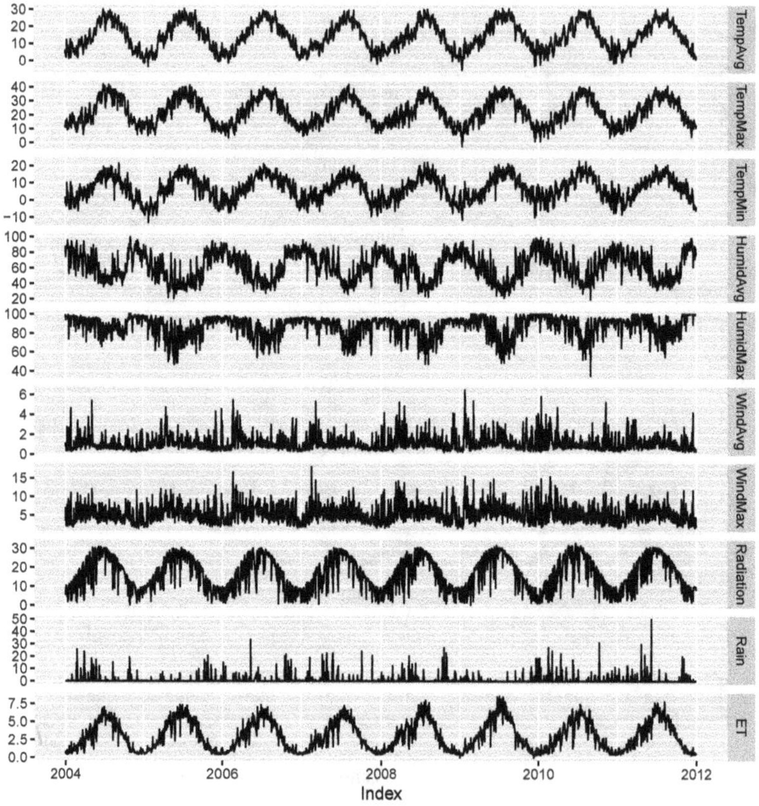

FIGURE 3.2: Time plot of the collection of meteorological time series of the Aranjuez station (ggplot2 version).

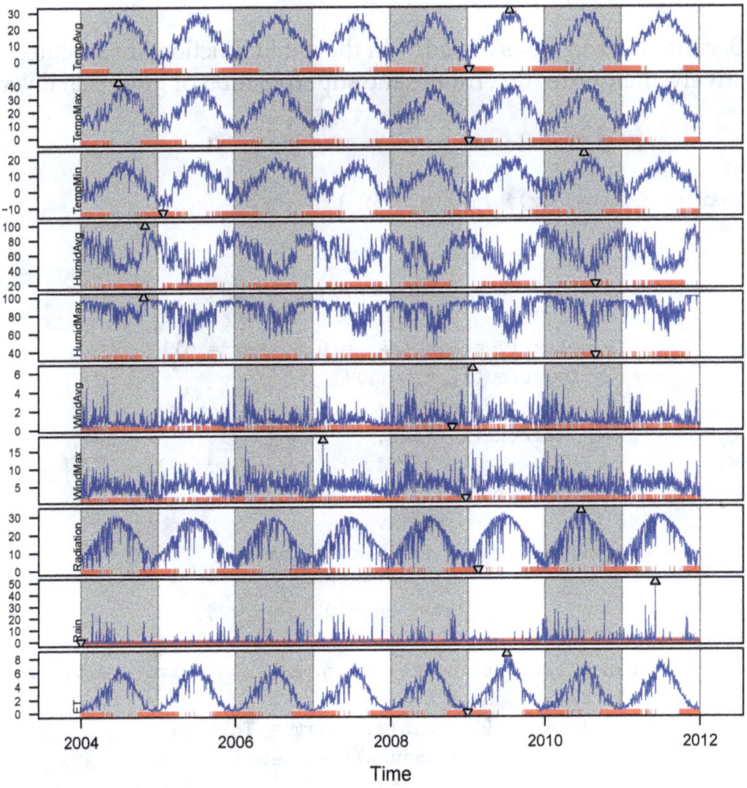

FIGURE 3.3: Enhanced time plot of the collection of meteorological time series of the Aranjuez station.

- The label of each time series is displayed with text inside each panel instead of using the strips mechanism. The `panel.text` prints the name of each variable with the aid of `panel.number` (line 21).

- The maxima and minima are highlighted with small blue triangles (lines 25 and 28, respectively).

Because the functions included in the panel function are executed consecutively, their order determines the superposition of graphical layers.

```
1   ## Auxiliary function to extract the year value of a POSIXct time
2   ## index
3   Year <- function(x)format(x, "%Y")
4
5   xyplot(aranjuez,
6          layout = c(1, ncol(aranjuez)),
7          strip = FALSE,
8          scales = list(y = list(cex = 0.6, rot = 0)),
9          panel = function(x, y, ...){
10             ## Alternation of years
11             panel.xblocks(x, Year,
12                           col = c("lightgray", "white"),
13                           border = "darkgray")
14             ## Values under the average highlighted with red regions
15             panel.xblocks(x, y < mean(y, na.rm = TRUE),
16                           col = "indianred1",
17                           height = unit(0.1, 'npc'))
18             ## Time series
19             panel.lines(x, y, col = 'royalblue4', lwd = 0.5, ...)
20             ## Label of each time series
21             panel.text(x[1], min(y, na.rm = TRUE),
22                        names(aranjuez)[panel.number()],
23                        cex = 0.6, adj = c(0, 0), srt = 90, ...)
24             ## Triangles to point the maxima and minima
25             idxMax <- which.max(y)
26             panel.points(x[idxMax], y[idxMax],
27                          col = 'black', fill = 'lightblue', pch = 24)
28             idxMin <- which.min(y)
29             panel.points(x[idxMin], y[idxMin],
30                          col = 'black', fill = 'lightblue', pch = 25)
31          })
```

There is no equivalent `panel.xblocks` function that can be used with ggplot2. Therefore, the ggplot2 version must explicitly compute the corresponding bands (years and regions below the average values):

- The first step in working with `ggplot` is to transform the zoo object into a `data.frame` in long format. `fortify` returns a `data.frame` with three columns: the time `Index`, a factor indicating the `Series`, and the corresponding `Value`.

```
timeIdx <- index(aranjuez)

aranjuezLong <- fortify(aranjuez, melt = TRUE)

summary(aranjuezLong)
```

```
     Index              Series          Value
Min.   :2004-01-01  TempAvg : 2898  Min.   :-12.98
1st Qu.:2005-12-29  TempMax : 2898  1st Qu.:  2.01
Median :2008-01-09  TempMin : 2898  Median :  8.62
Mean   :2008-01-03  HumidAvg: 2898  Mean   : 22.05
3rd Qu.:2010-01-03  HumidMax: 2898  3rd Qu.: 27.62
Max.   :2011-12-31  WindAvg : 2898  Max.   :100.00
                    (Other) :11592  NA's   :37
```

- The bands of values below the average can be easily extracted with `scale` because these regions are negative when the `data.frame` is centered.

```
## Values below mean are negative after being centered
scaled <- fortify(scale(aranjuez, scale = FALSE), melt = TRUE)
## The 'scaled' column is the result of the centering.
## The new 'Value' column store the original values.
scaled <- transform(scaled, scaled = Value,
                Value = aranjuezLong$Value)
underIdx <- which(scaled$scaled <= 0)
## 'under' is the subset of values below the average
under <- scaled[underIdx,]
```

- The years bands are defined with the function `endpoints` from the xts package:

```
library("xts")
ep <- endpoints(timeIdx, on = 'years')
ep <- ep[-1]
N <- length(ep)
## 'tsp' is start and 'tep' is the end of each band. One of each
    two
## years are selected.
tep <- timeIdx[ep[seq(1, N, 2)] + 1]
tsp <- timeIdx[ep[seq(2, N, 2)]]
```

- The minima and maxima points of each variable are extracted with `apply`:

```
minIdx <- timeIdx[apply(aranjuez, 2, which.min)]
minVals <- apply(aranjuez, 2, min, na.rm = TRUE)
mins <- data.frame(Index = minIdx,
                Value = minVals,
                Series = names(aranjuez))

maxIdx <- timeIdx[apply(aranjuez, 2, which.max)]
maxVals <- apply(aranjuez, 2, max, na.rm = TRUE)
maxs <- data.frame(Index = maxIdx,
                Value = maxVals,
                Series = names(aranjuez))
```

- With ggplot we define the canvas, and the layers of information are added successively:

```
ggplot(data = aranjuezLong, aes(Index, Value)) +
    ## Time series of each variable
    geom_line(colour = "royalblue4", lwd = 0.5) +
    ## Year bands
    annotate('rect',
            xmin = tsp, xmax = tep,
            ymin = -Inf, ymax = Inf,
            alpha = 0.4) +
    ## Values below average
    geom_rug(data = under,
```

```
           sides = 'b', col = 'indianred1') +
## Minima
geom_point(data = mins, pch = 25) +
## Maxima
geom_point(data = maxs, pch = 24) +
## Axis labels and theme definition
labs(x = 'Time', y = NULL) +
theme_bw() +
## Each series is displayed in a different panel with an
## independent y scale
facet_free()
```

Some messages from Figure 3.3:

- The radiation, temperature, and evapotranspiration are quasi-periodic and are almost synchronized between them. Their local maxima appear in the summer and the local minima in the winter. Obviously, the summer values are higher than the average.

- The average humidity varies in opposition to the temperature and radiation cycle, with local maxima located during winter.

- The average and maximum wind speed and rainfall vary in a more erratic way and do not show the evident periodic behavior of the radiation and temperature.

- The rainfall is different from year to year. The remaining variables do not show variations between years.

- The fluctuations of solar radiation are more apparent than the temperature fluctuations. There is hardly any day with temperatures below the average value during summer, while it is not difficult to find days with radiation below the average during this season.

3.2   Time Series of Variables with the Same Scale

As an example of time series of variables with the same scale, we will use measurements of solar radiation from different meteorological stations.

The first attempt to display this multivariate time series makes use of the `xyplot.zoo` method. The objective of this graphic is to display the behavior of the collection as a whole: the series are superposed in the same panel (`superpose=TRUE`) without a legend (`auto.key=FALSE`), using thin lines and partial transparency[1]. Transparency softens overplotting

---

[1] A similar result can be obtained with `autoplot` using `facets=NULL`.

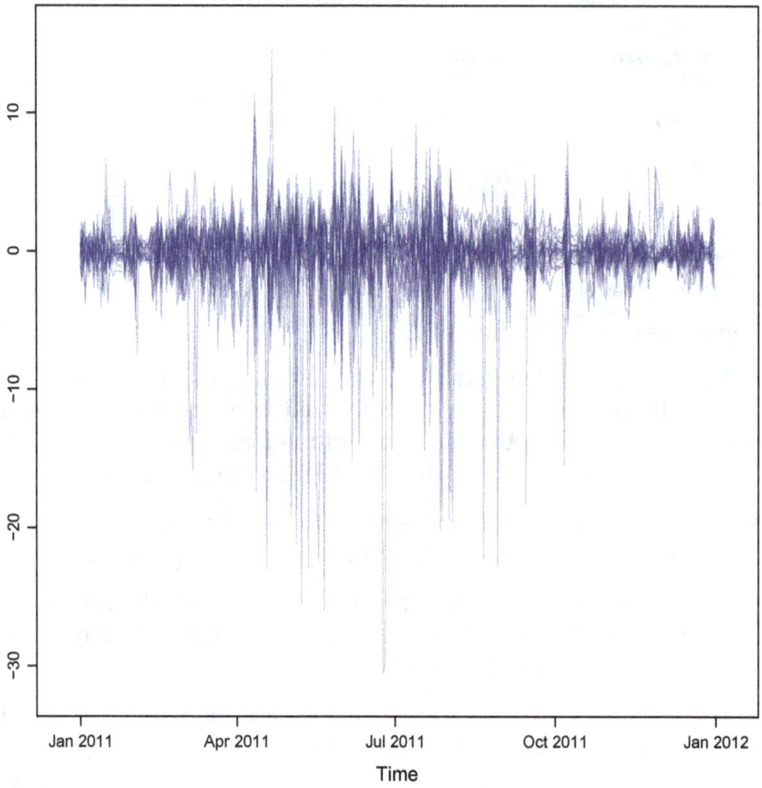

FIGURE 3.4: Time plot of the variations around the time average of solar radiation measurements from the meteorological stations of Navarra.

problems and reveals density clusters because regions with more overlapping lines are darker. Figure 3.4 displays the variations around the time average (avRad).

```
load('data/TimeSeries/navarra.RData')
```

```
avRad <- zoo(rowMeans(navarra, na.rm = 1), index(navarra))
pNavarra <- xyplot(navarra - avRad,
            superpose = TRUE, auto.key = FALSE,
            lwd = 0.5, alpha = 0.3, col = 'midnightblue')
pNavarra
```

This result can be improved with different methods: the cut-and-stack method and the horizon graph with `horizonplot`.

3.2.1 Aspect Ratio and Rate of Change

When a graphic is intended to inform about the rate of change, special attention must be paid to the aspect ratio of the graph, defined as the ratio of the height to the width of the graphical window. Cleveland analyzed the importance of the aspect ratio for judging rate of change (Cleveland and McGill 1984). He concluded that we visually decode the information about the relative local rate of change of one variable with another by comparing the orientations of the local line segments that compose the polylines. The recommendation is to choose the aspect ratio so that the absolute values of the orientations of the segments are centered on 45° (banking to 45°).

The problem with banking to 45° is that the resulting aspect ratio is frequently too small. A suitable solution to minimize wasted space is the cut-and-stack method. The `xyplot.ts` method implements this solution with the combination of the arguments `aspect` and `cut`. The version of Figure 3.4 using banking to 45° and the cut-and-stack method is produced with (Figure 3.5):

```
xyplot(navarra - avRad,
       aspect = 'xy', cut = list(n = 3, overlap = 0.1),
       strip = FALSE,
       superpose = TRUE, auto.key = FALSE,
       lwd = 0.5, alpha = 0.3, col = 'midnightblue')
```

3.2.2 The Horizon Graph

The horizon graph is useful in examining how a large number of series changes over time and does so in a way that allows both comparisons between the individual time series and independent analysis of each series. Moreover, extraordinary behaviors and predominant patterns are easily distinguished (Heer, Kong, and Agrawala 2009; Few 2008).

This graph displays several stacked series collapsing the y-axis to free vertical space:

- Positive and negative values share the same vertical space. Negative values are inverted and placed above the reference line. Sign is encoded using different hues (positive values in blue and negative values in red).

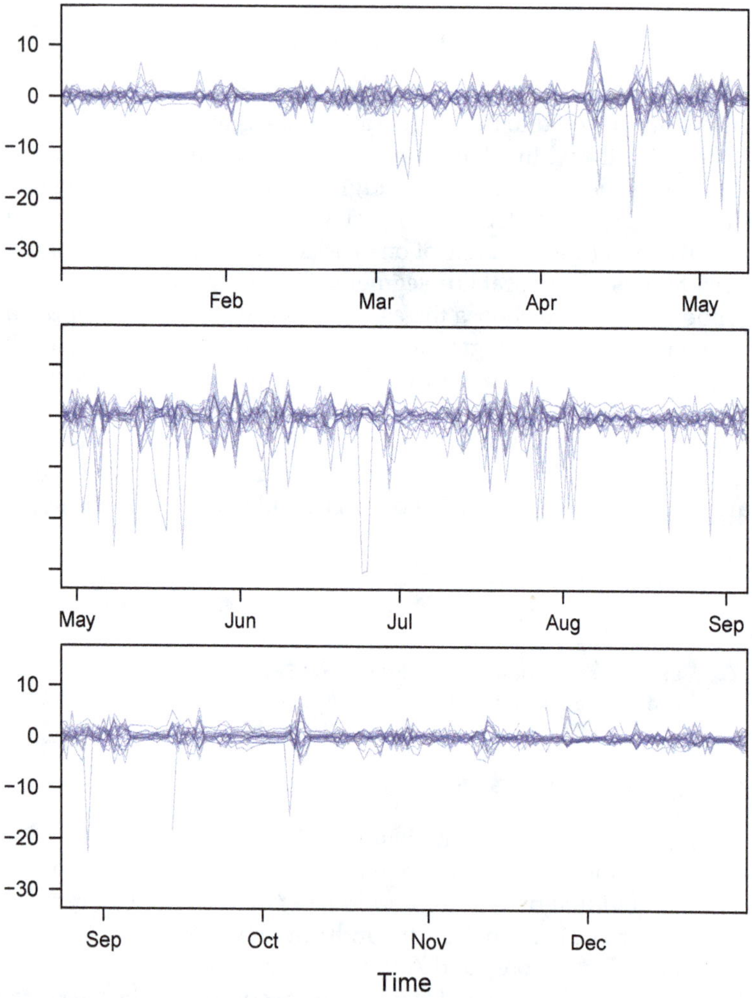

FIGURE 3.5: Cut-and-stack plot with banking to 45°.

- Differences in magnitude are displayed as differences in color intensity (darker colors for greater differences).

- The color bands share the same baseline and are superposed, with darker bands in front of the lighter ones.

Because the panels share the same design structure, once this technique is understood, it is easy to establish comparisons or spot extraordinary events. This method is what Tufte described as small multiples (Tufte 1990).

Figure 3.6 displays the variations of solar radiation around the time average with a horizon graph using a row for each time series. In the code we choose origin=0 and leave the argument horizonscale undefined (default). With this combination, each panel has different scales, and the colors in each panel represent deviations from the origin. This is depicted in the color key with the Δ_i symbol, where the subscript i denotes the existence of multiple panels with different scales.

```
horizonplot(navarra - avRad,
           layout = c(1, ncol(navarra)),
           origin = 0, ## Deviations in each panel are calculated
                      ## from this value
           colorkey = TRUE,
           col.regions = brewer.pal(6, "RdBu"))
```

Figure 3.6 allows several questions to be answered:

- Which stations consistently measure above and below the average?

- Which stations resemble more closely the average time series?

- Which stations show erratic and uniform behavior?

- In each of the stations, is there any day with extraordinary measurements?

- Which part of the year is associated with more intense absolute fluctuations across the set of stations?

3.2.3 Time Graph of the Differences between a Time Series and a Reference

The horizon graph is also useful in revealing the differences between a univariate time series and another reference. For example, we might be

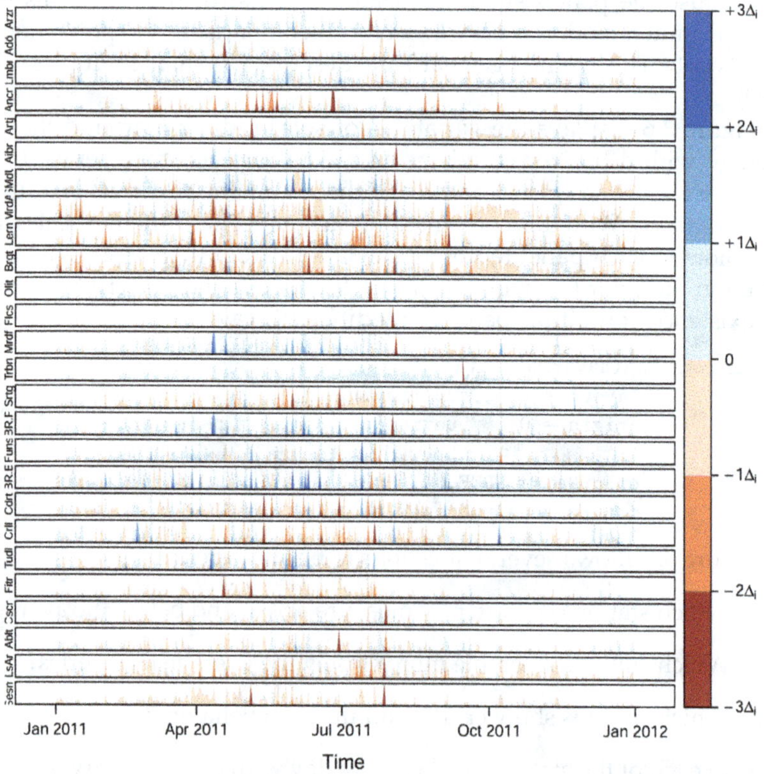

FIGURE 3.6: Horizon plot of variations around the time average of solar radiation measurements from the meteorological stations of Navarra.

interested in the departure of the observed temperature from the long-term average, or in other words, the temperature change over time.

Let's illustrate this approach with the time series of daily average temperatures measured at the meteorological station of Aranjuez. The reference is the long-term daily average calculated with ave.

```
Ta <- aranjuez$TempAvg
timeIndex <- index(aranjuez)
longTa <- ave(Ta, format(timeIndex, '%j'))
diffTa <- (Ta - longTa)
```

The temperature time series, the long-term average, and the differences between them can be displayed with the xyplot method, now using screens to use a different panel for the differences time series (Figure 3.7).

```
xyplot(cbind(Ta, longTa, diffTa),
       col = c('darkgray', 'red', 'midnightblue'),
       superpose = TRUE, auto.key = list(space = 'right'),
       screens = c(rep('Average Temperature', 2), 'Differences'))
```

The horizon graph is better suited for displaying the differences. The next code again uses the cut-and-stack method (Figure 3.5) to distinguish between years. Figure 3.8 shows that 2004 started clearly above the average, while 2005 and 2009 did the contrary. The year 2007 was frequently below the long-term average, but 2011 was more similar to that reference.

```
years <- unique(format(timeIndex, '%Y'))

horizonplot(diffTa, cut = list(n = 8, overlap = 0),
            colorkey = TRUE,
            col.regions = brewer.pal(6, "RdBu"),
            layout = c(1, 8),
            scales = list(draw = FALSE, y = list(relation = 'same'))
            ,
            origin = 0, strip.left = FALSE) +
    layer(grid.text(years[panel.number()], x = 0, y = 0.1,
                    gp = gpar(cex = 0.8),
                    just = "left"))
```

A different approach to display this information is to produce a level plot displaying the time series using parts of its time index as independent and conditioning variables[2]. The following code displays the differences

---

[2]This approach was inspired by the **strip** function of the `metvurst` package https://metvurst.wordpress.com/2013/03/04/visualising-large-amounts-of-hourly-environmental-data-2/

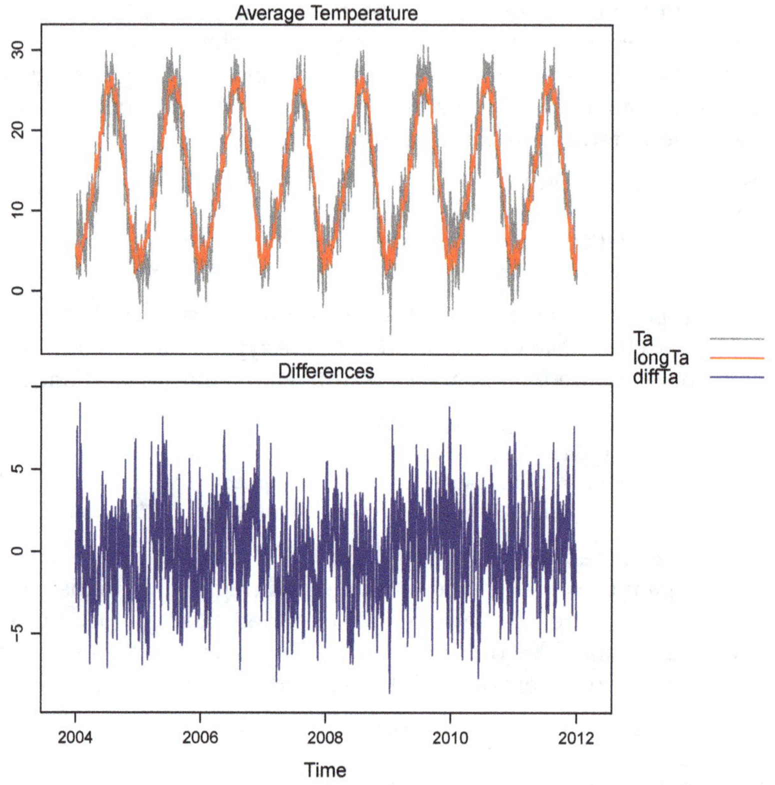

FIGURE 3.7: Daily temperature time series.

with the day of the month on the horizontal axis and the year on the vertical axis, with a different panel for each month number. Therefore, each cell of Figure 3.9 corresponds to a certain day of the time series. If you compare this figure with the horizon plot, you will find the same previous findings but revealed now in more detail. On the other hand, while the horizon plot of Figure 3.8 clearly displays the yearly evolution, the combination of variables of the level plot focuses on the comparison between years in a certain month.

```
year <- function(x)as.numeric(format(x, '%Y'))
day <- function(x)as.numeric(format(x, '%d'))
month <- function(x)as.numeric(format(x, '%m'))
```

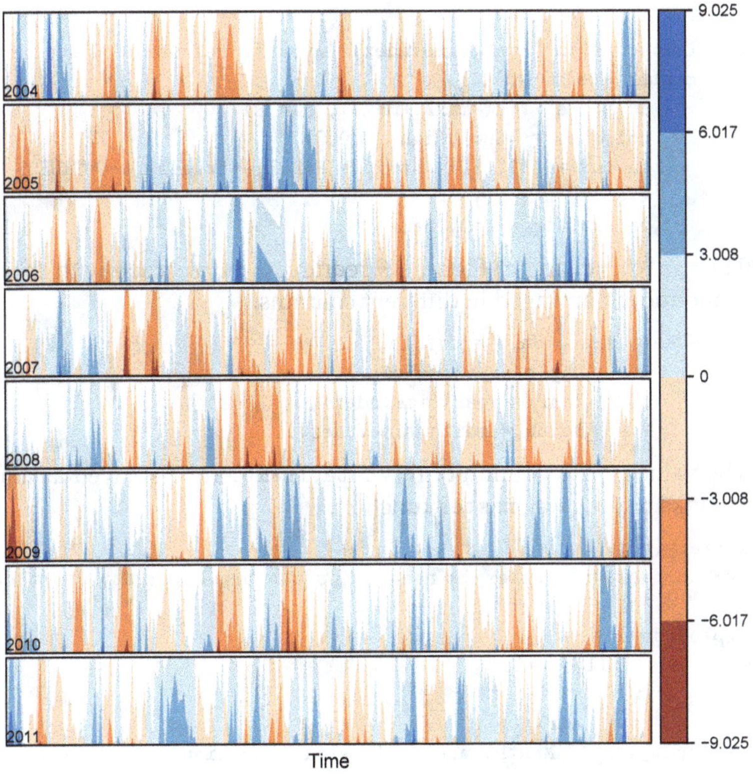

FIGURE 3.8: Horizon graph displaying differences between a daily temperature time series and its long-term average.

```
myTheme <- modifyList(custom.theme(region = brewer.pal(9, 'RdBu')
    ),
                    list(
                        strip.background = list(col = 'gray'),
                        panel.background = list(col = 'gray')))

maxZ <- max(abs(diffTa))

levelplot(diffTa ~ day(timeIndex) * year(timeIndex) | factor(
    month(timeIndex)),
        at = pretty(c(-maxZ, maxZ), n = 8),
        colorkey = list(height = 0.3),
        layout = c(1, 12), strip = FALSE, strip.left = TRUE,
        xlab = 'Day', ylab = 'Month',
        par.settings = myTheme)
```

The ggplot version of Figure 3.9 requires a data.frame with the day, year, and month arranged in different columns.

```
df <- data.frame(Vals = diffTa,
                Day = day(timeIndex),
                Year = year(timeIndex),
                Month = month(timeIndex))
```

The values (Vals column of this data.frame) are displayed as a level plot thanks to the geom_raster function.

```
library("scales")
## The packages scales is needed for the breaks_pretty function.

ggplot(data = df,
    aes(fill = Vals,
        x = Day,
        y = Year)) +
    facet_wrap(~ Month, ncol = 1, strip.position = 'left') +
    scale_y_continuous(breaks = breaks_pretty()) +
    scale_fill_distiller(palette = 'RdBu', direction = 1) +
    geom_raster() +
    theme(panel.grid.major = element_blank(),
        panel.grid.minor = element_blank())
```

3.3   Stacked Graphs

If the variables of a multivariate time series can be summed to produce a meaningful global variable, they may be better displayed with stacked

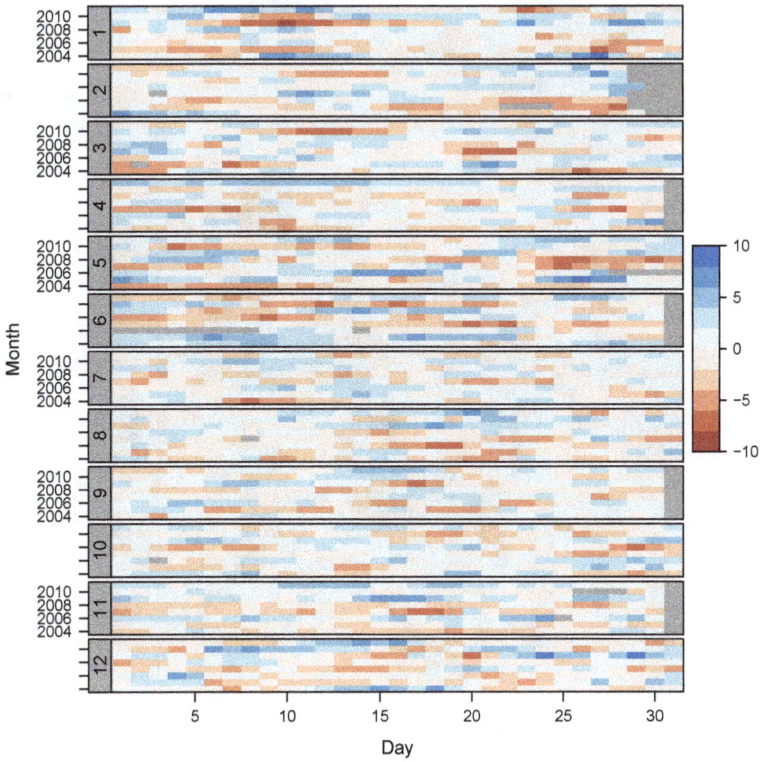

FIGURE 3.9: Level plot of differences between a daily temperature time series and its long-term average.

graphs. For example, the information on unemployment in the United States provides data of unemployed persons by industry and class of workers and can be summed to give a total unemployment time series.

```
load('data/TimeSeries/unemployUSA.RData')
```

The time series of unemployment can be directly displayed with the xyplot.zoo method (Figure 3.10).

```
xyplot(unemployUSA,
    superpose = TRUE,
    par.settings = custom.theme,
    auto.key = list(space = 'right'))
```

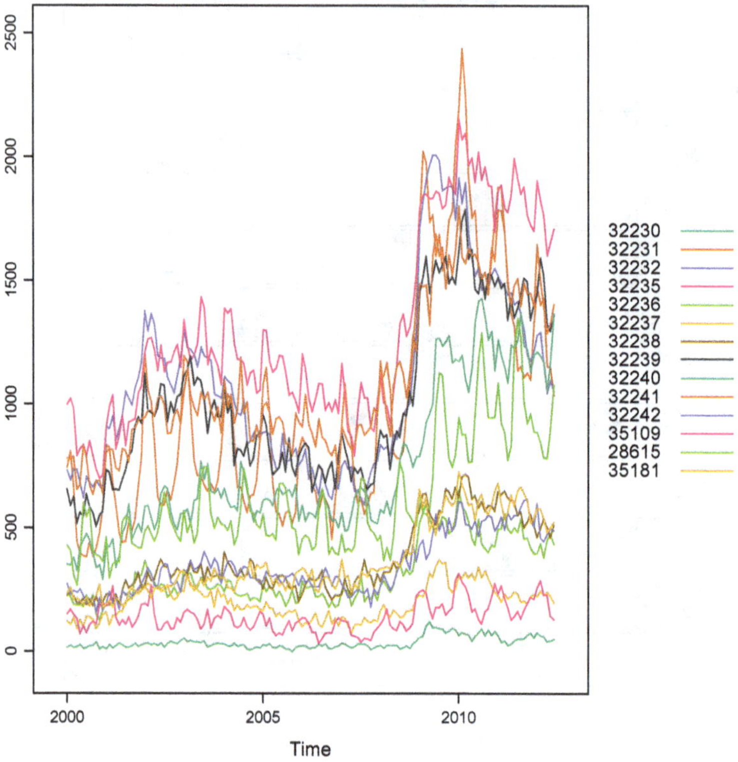

FIGURE 3.10: Time series of unemployment with xyplot.

This graphical output is not very useful: the legend includes too many items; the vertical scale is dominated by the largest series, with several series buried in the lower part of the scale; and the trend, variations, and structure of the total and individual contributions cannot be deduced from this graph.

A partial improvement is to display the multivariate time series as a set of stacked colored polygons to follow the macro/micro principle proposed by Tufte (Tufte 1990): Show a collection of individual time series and also display their sum. A traditional stacked graph is easily obtained with geom_area (Figure 3.11):

```
autoplot(unemployUSA, facets = NULL) +
    geom_area(aes(fill = Series)) +
    scale_x_yearmon()
```

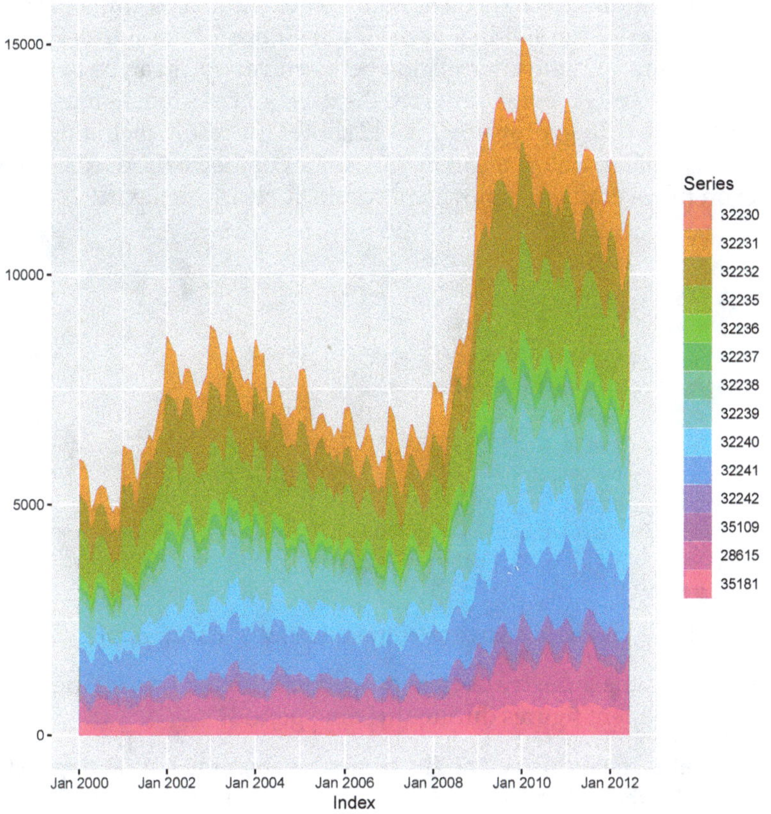

FIGURE 3.11: Time series of unemployment with stacked areas using geom_area.

Traditional stacked graphs have their bottom on the x-axis, which makes the overall height at each point easy to estimate. On the other hand, with this layout, individual layers may be difficult to distinguish. The *ThemeRiver* (Havre et al. 2002) (also named *streamgraph* in (Byron and Wattenberg 2008)) provides an innovative layout method in which layers are symmetrical around the x-axis at their center. At a glance, the pattern of the global sum and individual variables, their contribution to conforming the global sum, and the interrelation between variables can be perceived.

The package `ggstream`[3] provides the `geom_stream` function to produce streamgraphs with ggplot. Figure 3.12 shows the result, including a vertical line to indicate one of the main milestones of the financial crisis, whose effect on the overall unemployment results is clearly evident.

```
library("ggstream")

sep2008 <- as.numeric(as.yearmon('2008-09'))

autoplot(unemployUSA, facets = NULL) +
  geom_stream(aes(fill = Series),
              color = "black",
              lwd = 0.25,
              bw = 0.5) +
  scale_x_yearmon() +
  geom_vline(xintercept = sep2008,
             lwd = 0.25, color = "gray50") +
  theme_bw()
```

This figure can help answer several questions. For example:

- What is the industry or class of worker with the lowest/highest unemployment figures during this time period?

- What is the industry or class of worker with the lowest/highest unemployment increases due to the financial crisis?

- There are a number of local maxima and minima of the total unemployment numbers. Are all the classes contributing to the maxima/minima? Do all the classes exhibit the same fluctuation behavior as the global evolution?

There is no solution available for the `lattice` package. Therefore, I have defined a panel and prepanel functions[4] to implement a ThemeRiver

---

[3]https://github.com/davidsjoberg/ggstream
[4]The code of these panel and prepanel functions is explained in Section 3.3.1.

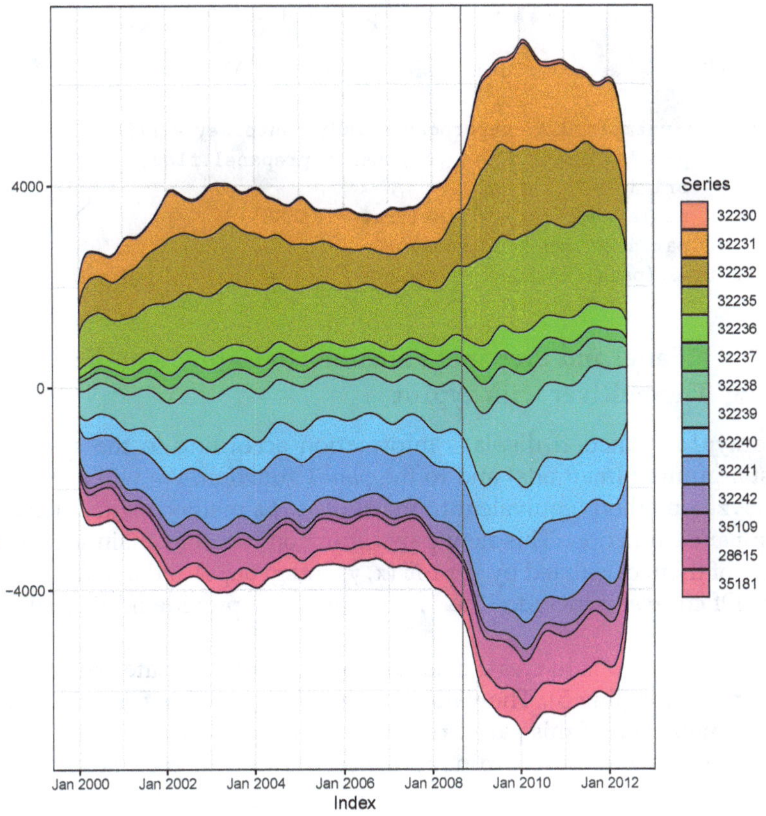

FIGURE 3.12: Streamgraph of unemployment in the United States.

with `xyplot`. These functions are fully explained in the next section. The result is displayed in Figure 3.13.

```
library("colorspace")
## We will use a qualitative palette from colorspace
nCols <- ncol(unemployUSA)
pal <- rainbow_hcl(nCols, c = 70, l = 75, start = 30, end = 300)
myTheme <- custom.theme(fill = pal, lwd = 0.2)

sep2008 <- as.numeric(as.yearmon('2008-09'))

xyplot(unemployUSA, superpose = TRUE, auto.key = FALSE,
       panel = panel.flow, prepanel = prepanel.flow,
       origin = 'themeRiver',
       scales = list(y = list(draw = FALSE)),
       par.settings = myTheme) +
    layer(panel.abline(v = sep2008, col = 'gray', lwd = 0.7))
```

3.3.1 ✎Panel and Prepanel Functions to Implement the ThemeRiver with `xyplot`

The `xyplot` function displays information according to the class of its first argument (methods) and to the `panel` function. We will use the `xyplot.zoo` method (equivalent to the `xyplot.ts` method) with a new custom panel function. This new panel function has four main arguments, three of them calculated by `xyplot` (x, y, and groups) and a new one, `origin`. Of course, it includes the ... argument to provide additional arguments.

The first step is to create a `data.frame` with coordinates and with the groups factor (line 5). The value and number of the levels will be used in the main step of this `panel` function. With this `data.frame` we have to calculate the y and x coordinates for each group to get a stacked set of polygons.

This `data.frame` is in the *long* format, with a row for each observation, and where the group column identifies the variable. Thus, it must be transformed to the *wide* format, with a column for each variable. With the `unstack` function, a new `data.frame` is produced whose columns are defined according to the formula y ~ groups and with a row for each time position (line 10). The stack of polygons is the result of the cumulative sum of each row (line 20). The origin of this sum is defined with the corresponding `origin` argument: with `themeRiver`, the polygons are arranged in a symmetric way.

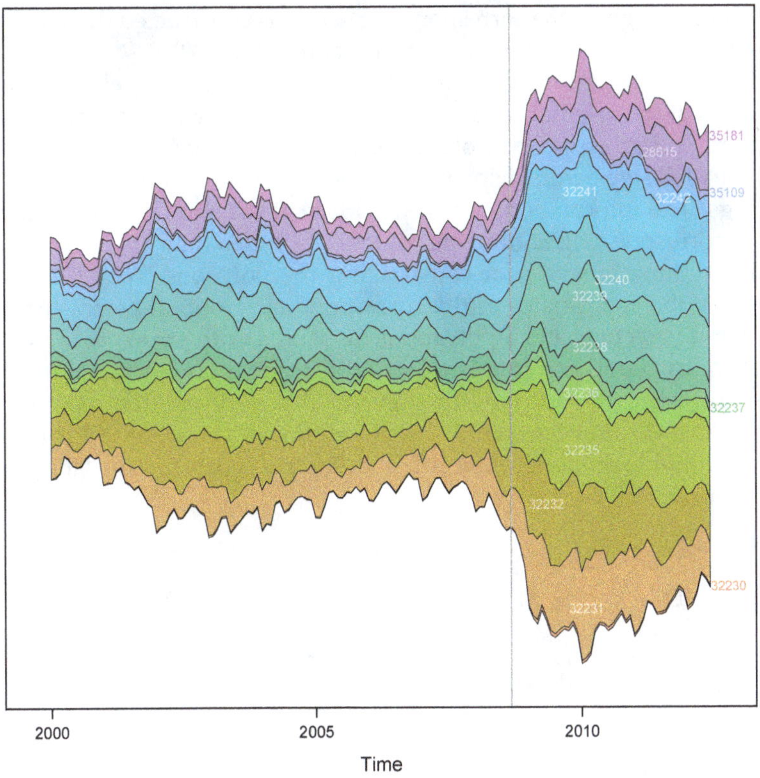

FIGURE 3.13: ThemeRiver of unemployment in the United States.

Each column of this matrix of cumulative sums defines the y coordinate of each variable (where the `origin` is now the first variable). The polygon of each variable is between this curve (`iCol+1`) and the one of the previous variable (`iCol`) (line 21). In order to get a closed polygon, the coordinates of the inferior limit are in reverse order. This new `data.frame` (Y) is in the *wide* format, but `xyplot` requires the information in the *long* format: the y coordinates of the polygons are extracted from the `values` column of the *long* version of this `data.frame` (line 26).

The x coordinates are produced in an easier way. Again, `unstack` produces a `data.frame` with a column for each variable and a row for each time position (line 29), but now, because the x coordinates are the same for the set of polygons, the corresponding vector is constructed directly using a combination of concatenation and repetition (line 30).

Finally, the groups vector is produced, repeating each element of the columns of the original `data.frame` (dat$groups) twice to account for the forward and reverse curves of the corresponding polygon (line 32).

The final step before displaying the polygons is to acquire the graphical settings. The information retrieved with `trellis.par.get` is transferred to the corresponding arguments of `panel.polygon` (line 35).

Everything is ready for constructing the polygons. With a `for` loop (line 41), the coordinates of the corresponding group are extracted from the x and y vectors, and a polygon is displayed with `panel.polygon`. The labels of each polygon (the `levels` of the original groups variable, groupLevels) are printed inside the polygon if there is enough room for the text (hChar>1) or at the right if the polygon is too small or if it is the first or last variable of the set (line 60). Both the polygons and the labels share the same color (`col[i]`).

```r
1   library("grid")
2
3   panel.flow <- function(x, y, groups, origin, ...)
4   {
5       dat <- data.frame(x = x, y = y, groups = groups)
6       nVars <- nlevels(groups)
7       groupLevels <- levels(groups)
8
9       ## From long to wide
10      yWide <- unstack(dat, y~groups)
11      ## Where are the maxima of each variable located? We will use
12      ## them to position labels.
13      idxMaxes <- apply(yWide, 2, which.max)
14
15      ##Origin calculated following Havr.eHetzler.ea2002
16      if (origin=='themeRiver') origin = -1/2*rowSums(yWide)
17      else origin = 0
18      yWide <- cbind(origin = origin, yWide)
19      ## Cumulative sums to define the polygon
20      yCumSum <- t(apply(yWide, 1, cumsum))
21      Y <- as.data.frame(sapply(seq_len(nVars),
22                          function(iCol)c(yCumSum[,iCol+1],
23                                  rev(yCumSum[,iCol]))))
24      names(Y) <- levels(groups)
25      ## Back to long format, since xyplot works that way
26      y <- stack(Y)$values
```

```
27
28    ## Similar but easier for x
29    xWide <- unstack(dat, x~groups)
30    x <- rep(c(xWide[,1], rev(xWide[,1])), nVars)
31    ## Groups repeated twice (upper and lower limits of the
          polygon)
32    groups <- rep(groups, each = 2)
33
34    ## Graphical parameters
35    superpose.polygon <- trellis.par.get("superpose.polygon")
36    col = superpose.polygon$col
37    border = superpose.polygon$border
38    lwd = superpose.polygon$lwd
39
40    ## Draw polygons
41    for (i in seq_len(nVars)){
42        xi <- x[groups==groupLevels[i]]
43        yi <- y[groups==groupLevels[i]]
44        panel.polygon(xi, yi, border = border,
45                      lwd = lwd, col = col[i])
46    }
47
48    ## Print labels
49    for (i in seq_len(nVars)){
50        xi <- x[groups==groupLevels[i]]
51        yi <- y[groups==groupLevels[i]]
52        N <- length(xi)/2
53        ## Height available for the label
54        h <- unit(yi[idxMaxes[i]], 'native') -
55            unit(yi[idxMaxes[i] + 2*(N-idxMaxes[i]) +1], 'native')
56        ##...converted to "char" units
57        hChar <- convertHeight(h, 'char', TRUE)
58        ## If there is enough space and we are not at the first or
59        ## last variable, then the label is printed inside the
              polygon.
60        if((hChar >= 1) && !(i %in% c(1, nVars))){
61            grid.text(groupLevels[i],
62                      xi[idxMaxes[i]],
63                      (yi[idxMaxes[i]] +
64                       yi[idxMaxes[i] + 2*(N-idxMaxes[i]) +1])/2,
65                      gp = gpar(col = 'white', alpha = 0.7, cex = 0.7)
                          ,
66                      default.units = 'native')
67        } else {
```

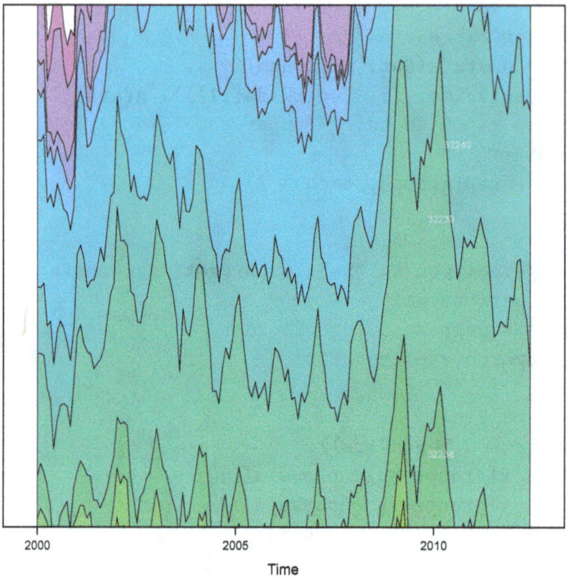

FIGURE 3.14: First attempt of ThemeRiver.

```
68        ## Elsewhere, the label is printed outside
69
70        grid.text(groupLevels[i],
71                  xi[N],
72                  (yi[N] + yi[N+1])/2,
73                  gp = gpar(col = col[i], cex = 0.7),
74                  just = 'left', default.units = 'native')
75    }
76  }
77 }
```

With this panel function, xyplot displays a set of stacked polygons corresponding to the multivariate time series (Figure 3.14). However, the graphical window is not large enough, and part of the polygons fall out of it. Why?

```
xyplot(unemployUSA, superpose = TRUE, auto.key = FALSE,
       panel = panel.flow, origin = 'themeRiver',
       par.settings = myTheme, cex = 0.4, offset = 0,
       scales = list(y = list(draw = FALSE)))
```

The problem is that `lattice` makes a preliminary estimate of the window size using a default `prepanel` function that is unaware of the internal calculations of our new `panel.flow` function. The solution is to define a new `prepanel.flow` function.

The input arguments and first lines are the same as in `panel.flow`. The output is a list whose elements are the limits for each axis (`xlim` and `ylim`, line 12) and the sequence of differences (`dx` and `dy`, line 14) that can be used for the aspect and banking calculations.

The limits of the x-axis are defined with the range of the time index, while the limits of the y-axis are calculated with the minimum of the first column of yCumSum (the origin line) and with the maximum of its last column (the upper line of the cumulative sum) (line 13).

```
1   prepanel.flow <- function(x, y, groups, origin,...)
2   {
3       dat <- data.frame(x = x, y = y, groups = groups)
4       nVars <- nlevels(groups)
5       groupLevels <- levels(groups)
6       yWide <- unstack(dat, y~groups)
7       if (origin=='themeRiver') origin = -1/2*rowSums(yWide)
8       else origin = 0
9       yWide <- cbind(origin = origin, yWide)
10      yCumSum <- t(apply(yWide, 1, cumsum))
11
12      list(xlim = range(x),
13          ylim = c(min(yCumSum[,1]), max(yCumSum[,nVars+1])),
14          dx = diff(x),
15          dy = diff(c(yCumSum[,-1])))
16  }
```

3.4 Interactive Graphics

This section describes the interactive alternatives of the static figures included in the previous sections with several packages: dygraphs, highcharter, plotly, and `streamgraph`. These packages are R interfaces to JavaScript libraries based on the `htmlwidgets` package.

3.4.1 Dygraphs

The dygraphs package is an interface to the dygraphs JavaScript library and provides facilities for charting time-series. It works automatically with

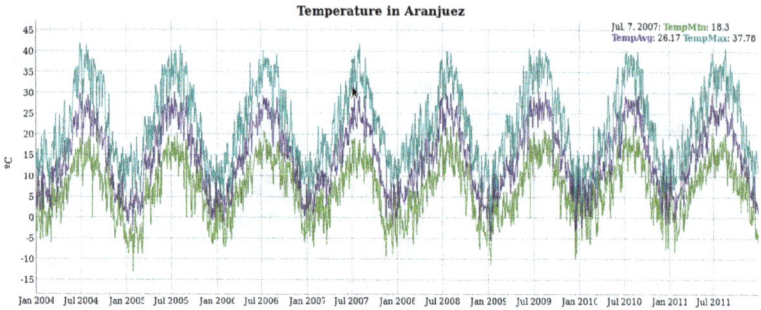

FIGURE 3.15: Snapshot of an interactive graphic produced with dygraphs.

xts time series objects or with objects that can be coerced to this class. The result is an interactive graph, where values are displayed according to the mouse position over the time series. Regions can be selected to zoom into a time period. Figure 3.15 is a snapshot of the interactive graph.

```
library("dygraphs")

dyTemp <- dygraph(aranjuez[, c("TempMin", "TempAvg", "TempMax")],
          main = "Temperature in Aranjuez",
          ylab = "ºC")
```

```
dyTemp
```

You can customize dygraphs by piping additional commands onto the original graphic. The function dyOptions provides several choices for the graphic, and the function dyHighlight configures options for data series mouse-over highlighting. For example, with the next code, the semi-transparency value of the non-selected lines is reduced and the width of the selected line is increased (Figure 3.16).

```
dyTemp %>%
    dyHighlight(highlightSeriesBackgroundAlpha = 0.2,
          highlightSeriesOpts = list(strokeWidth = 2))
```

An alternative approach to depict the upper and lower variables of this time series is with a shaded region. The dySeries function accepts a character vector of length 3 that specifies a set of input column names to use as the lower, value, and upper for a series with a shaded region around it (Figure 3.17).

FIGURE 3.16: Snapshot of a selection in an interactive graphic produced with dygraphs.

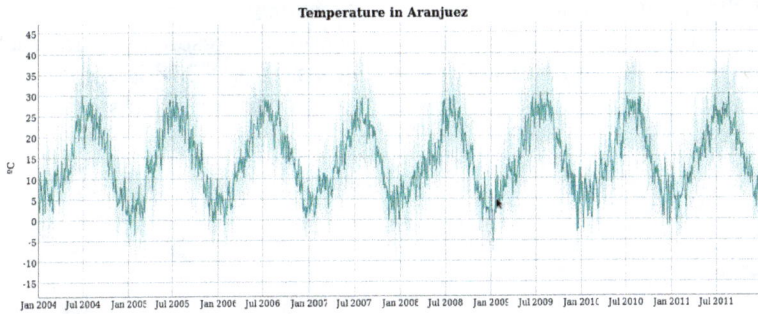

FIGURE 3.17: Shaded region between upper and lower values around a time series.

```
dygraph(aranjuez[, c("TempMin", "TempAvg", "TempMax")],
      main = "Temperature in Aranjuez",
      ylab = "ºC") %>%
   dySeries(c("TempMin", "TempAvg", "TempMax"),
          label = "Temperature")
```

3.4.2 Highcharter

The highcharter package is an interface to the highcharts JavaScript library, with a wide spectrum of graphics solutions. Displaying time series with this package can be achieved with the combination of the generic

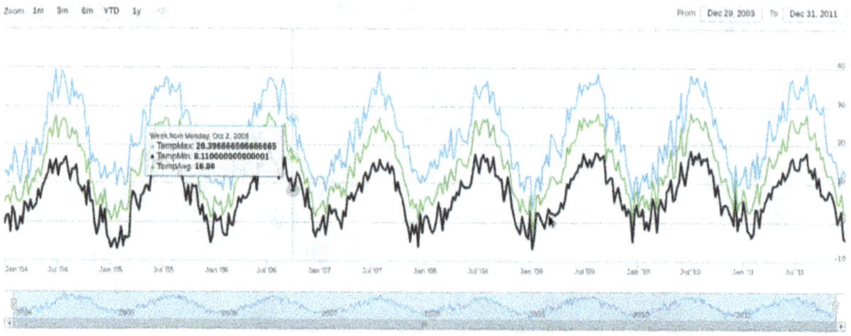

FIGURE 3.18: Snapshot of an interactive graphic produced with `highchar-ter`.

`highchart` function and several calls to the `hc_add_series_xts` function through the pipe `%>%` operator. Once again, the result is an interactive graph with selection and zoom capabilities. Figure 3.18 is a snapshot of the interactive graph, and Figure 3.19 is a snapshot of this same graph with zoom.

```
library("highcharter")
library("xts")

aranjuezXTS <- as.xts(aranjuez)

highchart(type = "stock") %>%
    hc_add_series(name = 'TempMax',
                       aranjuezXTS[, "TempMax"]) %>%
    hc_add_series(name = 'TempMin',
                       aranjuezXTS[, "TempMin"]) %>%
    hc_add_series(name = 'TempAvg',
                       aranjuezXTS[, "TempAvg"])
```

3.4.3 plotly

The `plotly` package is an interface to the `plotly` JavaScript library, also with a wide spectrum of graphics solutions. This package does not provide any function specifically focused on time series. Thus, the time series object has to be transformed into a `data.frame` including a column for the time index. If the `data.frame` is in *wide* format (one column per variable), each variable will be represented with a call to the `add_lines` function.

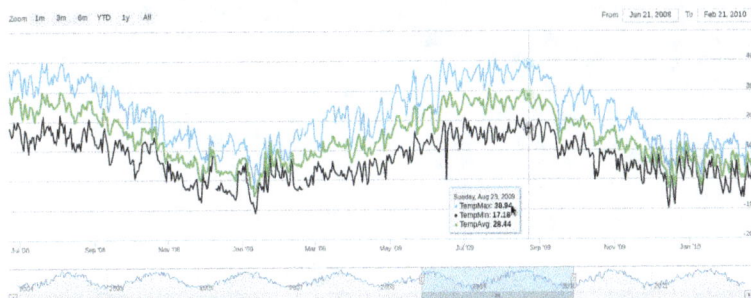

FIGURE 3.19: Snapshot of a zoom in an interactive graphic produced with `highcharter`.

However, if the `data.frame` is in *long* format (a column for values and a column for variable names), only one call to `add_lines` is required. The next code follows this approach using the combination of `fortify`, to convert the `zoo` object into a `data.frame`, and `melt`, to transform from wide to long format.

```
aranjuezDF <- fortify(aranjuez[,
                        c("TempMax",
                          "TempAvg",
                          "TempMin")],
                melt = TRUE)

summary(aranjuezDF)
```

```
     Index                Series              Value
 Min.   :2004-01-01   TempMax:2898    Min.    :-12.980
 1st Qu.:2005-12-29   TempAvg:2898    1st Qu.:  7.107
 Median :2008-01-09   TempMin:2898    Median : 13.560
 Mean   :2008-01-03                   Mean    : 14.617
 3rd Qu.:2010-01-03                   3rd Qu.: 21.670
 Max.   :2011-12-31                   Max.    : 41.910
                                      NA's    :10
```

Figure 3.20 is a snapshot of the interactive graphic produced with the generic function `plot_ly` connected with `add_lines` through the pipe operator, `%>%`.

49

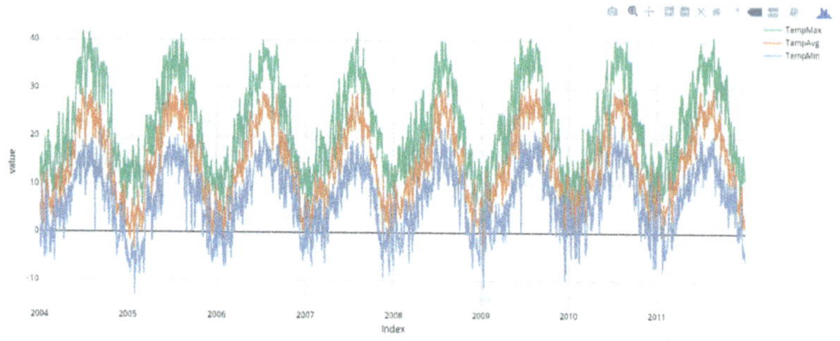

FIGURE 3.20: Snapshot of an interactive graphic produced with `plotly`.

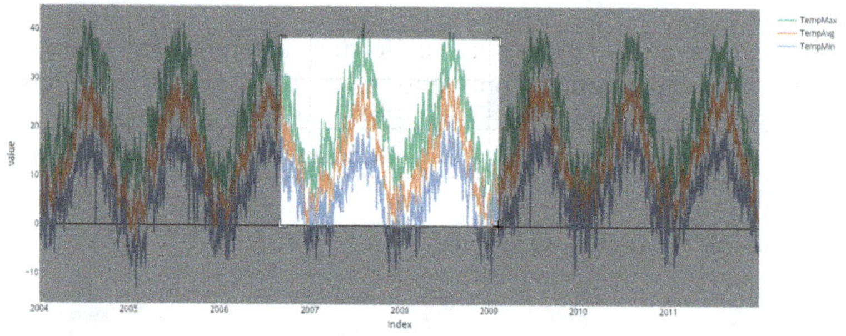

FIGURE 3.21: Snapshot of a zoom in an interactive graphic produced with `plotly`.

```
library("plotly")

plot_ly(aranjuezDF) %>%
    add_lines(x = ~ Index,
              y = ~ Value,
              color = ~ Series)
```

3.4.4 streamgraph

The `streamgraph` package[5] creates interactive stream graphs based on the `htmlwidgets` package and the `D3.js` JavaScript library.

---

[5]The `streamgraph` package, http://hrbrmstr.github.io/streamgraph/, is not available in CRAN. It can be installed using the `devtools` or the `remotes` package.

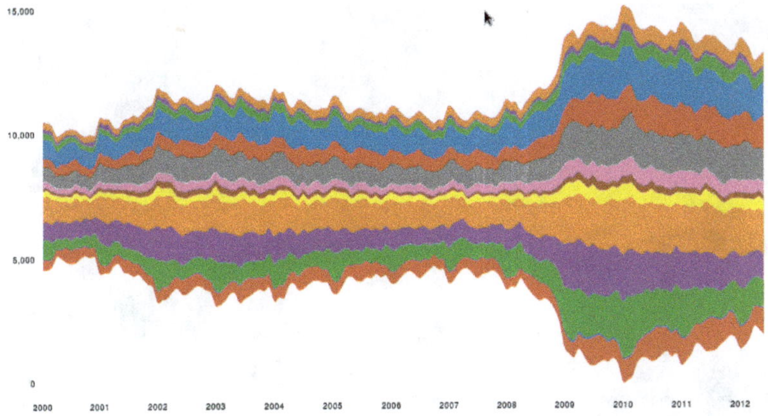

FIGURE 3.22: Streamgraph created with the `streamgraph` package, without selection.

```
## remotes::install_github("hrbrmstr/streamgraph")
library("streamgraph")
```

Its main function, `streamgraph`, requires a `data.frame` as the first argument. Besides, its three next arguments, `key`, `value`, and `date`, make this function a good candidate to work together with `fortify` and `melt`.

```
unemployDF <- fortify(unemployUSA, melt = TRUE)
## streamgraph does not work with the yearmon class
unemployDF$Index <- as.Date(unemployDF$Index)

head(unemployDF)
```

```
  Index Series Value
1 2000-01-01   32230    19
2 2000-02-01   32230    25
3 2000-03-01   32230    17
4 2000-04-01   32230    20
5 2000-05-01   32230    27
6 2000-06-01   32230    13
```

Figures 3.22 and 3.23 are snapshots of the interactive graphic created with the functions `streamgraph`, `sg_axis`, and `sg_fill_brewer`, connected through the pipe operator, `%>%`.

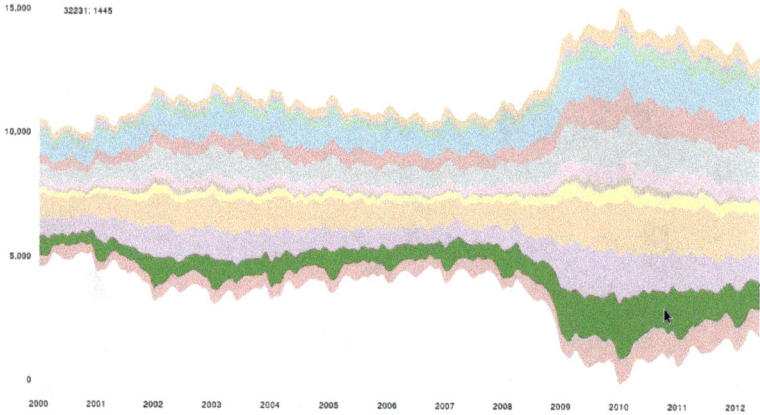

FIGURE 3.23: Streamgraph created with the `streamgraph` package, with a selection.

```
streamgraph(unemployDF,
          key = "Series",
          value = "Value",
          date = "Index") %>%
     sg_axis_x(1, "year", "%Y") %>%
     sg_fill_brewer("Set1")
```

Chapter 4

Time as a Conditioning or Grouping Variable

In Section 3.1 we learned to display the time evolution of multiple time series with different scales. But what if, instead of displaying the time evolution, we want to represent the relation between the variables? This chapter follows this approach: Section 4.1 proposes the scatterplot matrix solution using groups defined according to the time as a grouping variable; Section 4.2 produces an enhanced scatterplot with time as a conditioning variable using the small multiples technique; Section 4.1.1 includes a discussion about the hexagonal binning for large datasets.

Along with this chapter, these subjects are covered: scatterplot matrices, small multiples, hexagonal binning, and panel functions.

The most relevant packages in this chapter are: zoo for reading and arranging data as time series; GGally for creating the scatterplot matrix with ggplot2; hexbin for hexagonal binning; and reshape2 for converting data from the wide format to the long format.

DOI: 10.1201/9781003485384-4

4.1 Scatterplot Matrix: Time as a Grouping Variable

The scatterplot matrices are based on the technique of small multiples (Tufte 1990): small, thumbnail-sized representations of multiple images displayed all at once, which allows the reader to immediately, and in parallel, compare the inter-frame differences. A scatterplot matrix is a display of all pairwise bivariate scatterplots arranged in a $p \times p$ matrix for p variables. Each subplot shows the relation between the pair of variables at the intersection of the row and column indicated by the variable names in the diagonal panels (Friendly and Denis 2005).

This graphical tool is implemented in the splom function. The following code displays the relation between the set of meteorological variables using a sequential palette from the ColorBrewer catalog (RbBu, with black added to complete a twelve-color palette) to encode the month. The order of colors of this palette is chosen in order to display summer months with intense colors and to distinguish between the first and second half of the year with red and blue, respectively (Figure 4.1).

```
library("zoo")

load('data/TimeSeries/aranjuez.RData')
aranjuezDF <- as.data.frame(aranjuez)
aranjuezDF$Month <- format(index(aranjuez), '%m')

## Red-Blue palette with black added (12 colors)
colors <- c(brewer.pal(n = 11, 'RdBu'), '#000000')
## Rearrange according to months (darkest for summer)
colors <- colors[c(6:1, 12:7)]

splom(~ aranjuezDF[1:10],
      groups = aranjuezDF$Month,
      auto.key = list(space = 'right',
                      title = 'Month', cex.title = 1),
      pscale = 0, varname.cex = 0.7, xlab = '',
      par.settings = custom.theme(symbol = colors,
                                  pch = 19),
      cex = 0.3, alpha = 0.1)
```

A bit of interactivity can be added to this plot with the identification of some points. This task is easy with panel.link.splom. The points are selected via mouse clicks (and highlighted in green). Clicks other than left-clicks terminate the procedure. The output of this function is the index of chosen points.

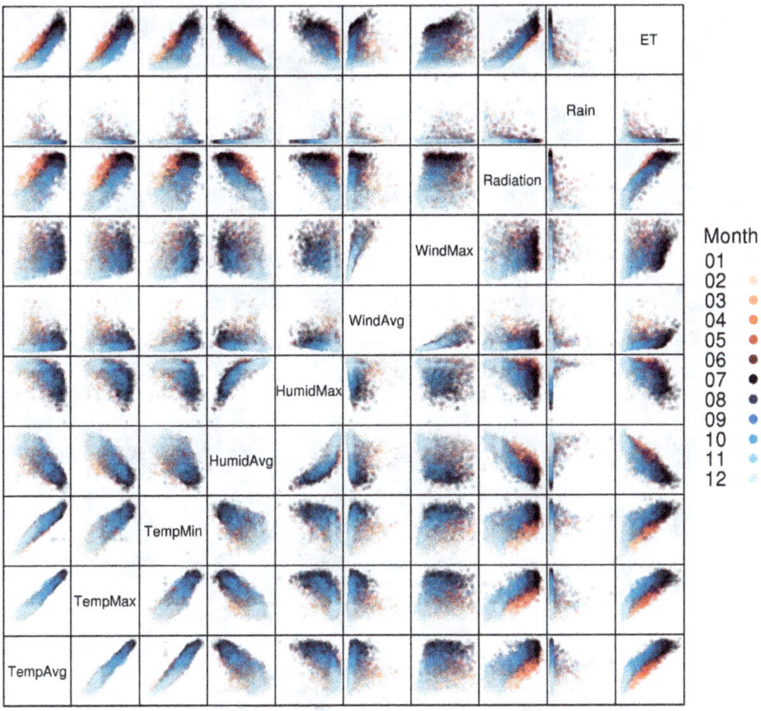

FIGURE 4.1: Scatter plot matrix of the collection of meteorological time series of the Aranjuez station (`lattice` version).

```
trellis.focus('panel', 1, 1)
idx <- panel.link.splom(pch = 13, cex = 0.6, col = 'green')
aranjuez[idx,]
```

The ggplot2 version of Figure 4.1 is produced thanks to the ggpairs function provided by the GGally package (Figure 4.2).

```
library("GGally")

ggpairs(aranjuezDF,
        columns = 1:10, ## Do not include "Month"
        upper = list(continuous = wrap("points", alpha = 0.7)),
        lower = list(continuous = wrap("points", alpha = 0.7)),
        diag = list(continuous = wrap("densityDiag", alpha = 0.7)),
        mapping = aes(colour = Month),
```

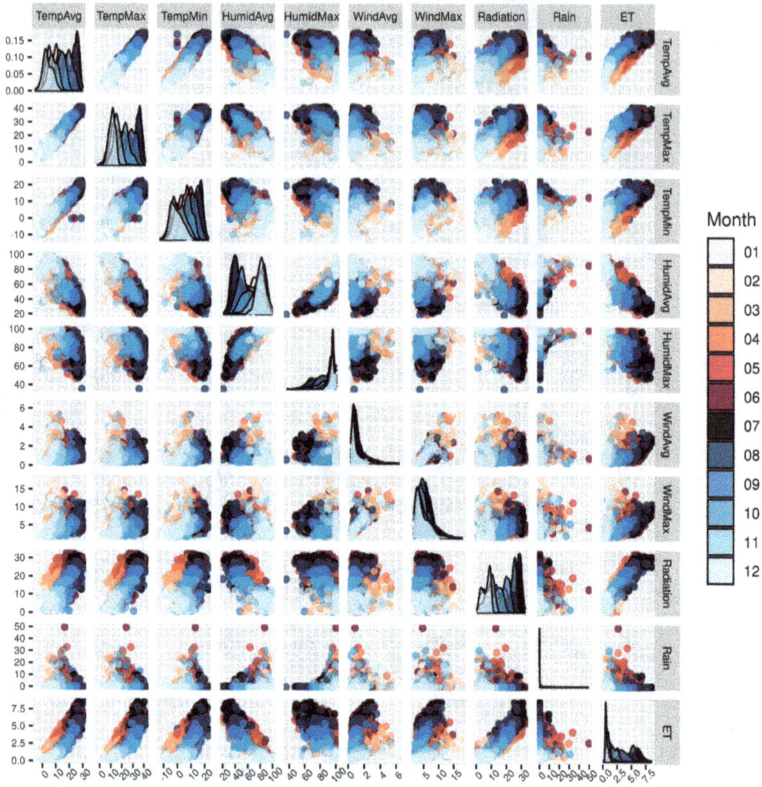

FIGURE 4.2: Scatter plot matrix of the collection of meteorological time series of the Aranjuez station (ggplot version).

```
        legend = 1) +
    scale_fill_manual(values = colors) +
    scale_colour_manual(values = colors) +
    theme(axis.text = element_text(size = 6),
        axis.text.x = element_text(angle = 45),
        strip.text = element_text(size = 7))
```

Let's explore Figures 4.1 and 4.2. For example,

- The highest values of ambient temperature (average, maximum, and minimum), solar radiation, and evapotranspiration can be found during the summer.

- These variables are almost linearly related. The relation between radiation and temperature is different during both halves of the year (red and blue regions can be easily distinguished).

- The humidity reaches its highest values during winter without appreciable differences between the first and second half of the year. The temperature and humidity may be related with an exponential function.

4.1.1 Hexagonal Binning

For large datasets, the display of a large number of points in a scatterplot produces hidden point density, long computation times, and slow displays. These problems can be circumvented with the estimation and representation of point densities. A common encoding uses gray scales, pseudo colors, or partial transparency. An improved scheme encodes density as the size of hexagon symbols inscribed within hexagonal binning regions (D. B. Carr et al. 1987).

The hexbin package (D. Carr, Lewin-Koh, and Maechler 2024) includes several functions for hexagonal binning. The panel.hexbinplot is a good substitute for the default panel function. In addition, our first attempt with splom can be improved with several modifications (Figure 4.3):

- The panels of the lower part of the matrix (lower.panel) will include a locally weighted scatterplot smoothing (loess) with panel.loess (line 12).

- The diagonal panels (diag.panel) will display the kernel density estimate of each variable (line 5). The density function computes this estimate. The result is adjusted to the panel limits (calculated with current.panel.limits). The kernel density is plotted with panel.lines and the diag.panel.splom function completes the content of each diagonal panel.

- The scale's ticks and labels are suppressed with pscale=0 (line 17)

- The point density is encoded with the default palette, LinGray (darker colors for high density values and lighter colors for almost empty regions, with a gradient of grays for intermediate values).

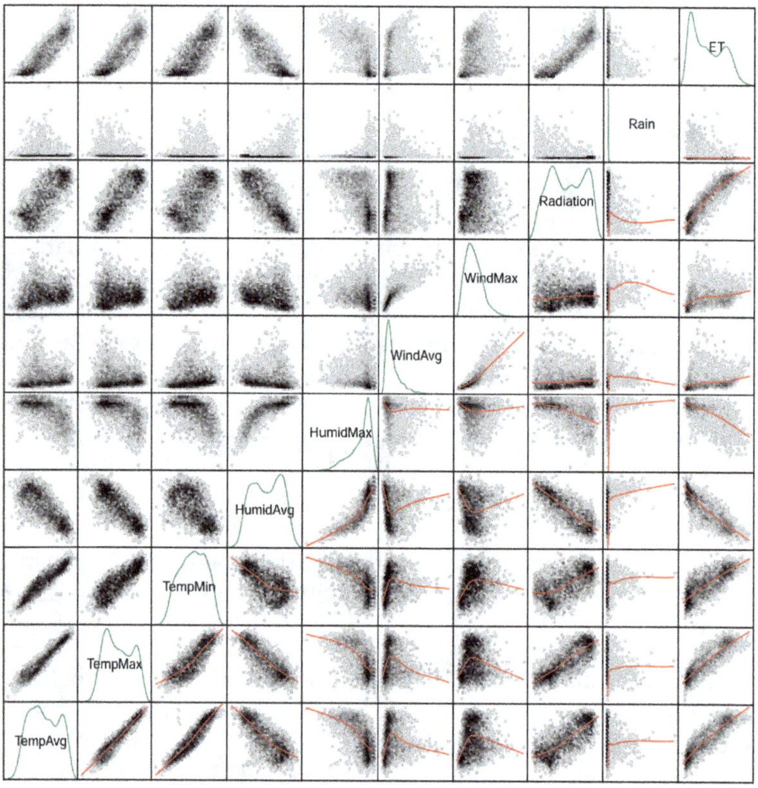

FIGURE 4.3: Scatterplot matrix of the collection of meteorological time series of the Aranjuez station using hexagonal binning.

```
1  library("hexbin")
2
3  splom(~as.data.frame(aranjuez),
4      panel = panel.hexbinplot,
5      diag.panel = function(x, ...){
6          yrng <- current.panel.limits()$ylim
7          d <- density(x, na.rm = TRUE)
8          d$y <- with(d, yrng[1] + 0.95 * diff(yrng) * y / max(y))
9          panel.lines(d)
10         diag.panel.splom(x, ...)
11     },
12     lower.panel = function(x, y, ...){
13         panel.hexbinplot(x, y, ...)
14         panel.loess(x, y, ..., col = 'red')
15     },
16     xlab = '',
17     pscale = 0,
18     varname.cex = 0.7)
```

A drawback of the matrix of scatterplots with hexagonal binning is that each panel is drawn independently, so it is impossible to compute a common color key for all of them. In other words, two cells with exactly the same color in different panels encode different point densities.

It is possible to display a reduced set of variables against another one and generate a common color key using the `hexbinplot` function. First, the dataset must be reshaped from the wide format (one column for each variable) to the long format (only one column for the temperature values with one row for each observation). This task is easily accomplished with the `melt` function included in the `reshape2` package.

```
library("reshape2")

aranjuezRshp <- melt(aranjuezDF,
                     measure.vars = c('TempMax',
                                      'TempAvg',
                                      'TempMin'),
                     variable.name = 'Statistic',
                     value.name = 'Temperature')

summary(aranjuezRshp)
```

```
   HumidAvg        HumidMax         WindAvg         WindMax
 Min.   :19.89   Min.   : 35.88   Min.   :0.250   Min.   : 1.550
```

```
 1st Qu.:47.04    1st Qu.: 81.60    1st Qu.:0.670    1st Qu.: 3.780
 Median :62.49    Median : 90.90    Median :0.920    Median : 5.030
 Mean   :62.11    Mean   : 87.20    Mean   :1.166    Mean   : 5.216
 3rd Qu.:77.30    3rd Qu.: 94.90    3rd Qu.:1.430    3rd Qu.: 6.540
 Max.   :99.50    Max.   :100.00    Max.   :6.450    Max.   :10.000
 NA's   :6        NA's   :33                         NA's   :345
    Radiation         Rain              ET               Month
 Min.   : 0.28    Min.   : 0.000    Min.   :0.000    Length:8694
 1st Qu.: 9.37    1st Qu.: 0.000    1st Qu.:1.160    Class :character
 Median :16.67    Median : 0.000    Median :2.750    Mode  :character
 Mean   :16.73    Mean   : 1.046    Mean   :3.088
 3rd Qu.:24.63    3rd Qu.: 0.200    3rd Qu.:4.923
 Max.   :32.74    Max.   :49.730    Max.   :8.560
                                    NA's   :42

    Statistic      Temperature
 TempMax:2898    Min.   :-12.980
 TempAvg:2898    1st Qu.:  7.107
 TempMin:2898    Median : 13.560
                 Mean   : 14.617
                 3rd Qu.: 21.670
                 Max.   : 41.910
                 NA's   :10
```

The hexbinplot displays this dataset with a different panel for each type of temperature (average, maximum, and minimum) but with a common color key encoding the point density (Figure 4.4). Now, two cells with the same color in different panels encode the same value.

```
hexbinplot(Radiation ~ Temperature | Statistic,
           data = aranjuezRshp,
           layout = c(1, 3)) +
    layer(panel.loess(..., col = 'red'))
```

The ggplot2 version is based on the geom_hex function (Figure 4.5).

```
ggplot(data = aranjuezRshp,
       aes(Temperature, Radiation)) +
    geom_hex() +
    stat_smooth(se = FALSE, method = 'loess', col = 'red') +
    facet_wrap(~ Statistic, ncol = 1) +
    theme_bw()
```

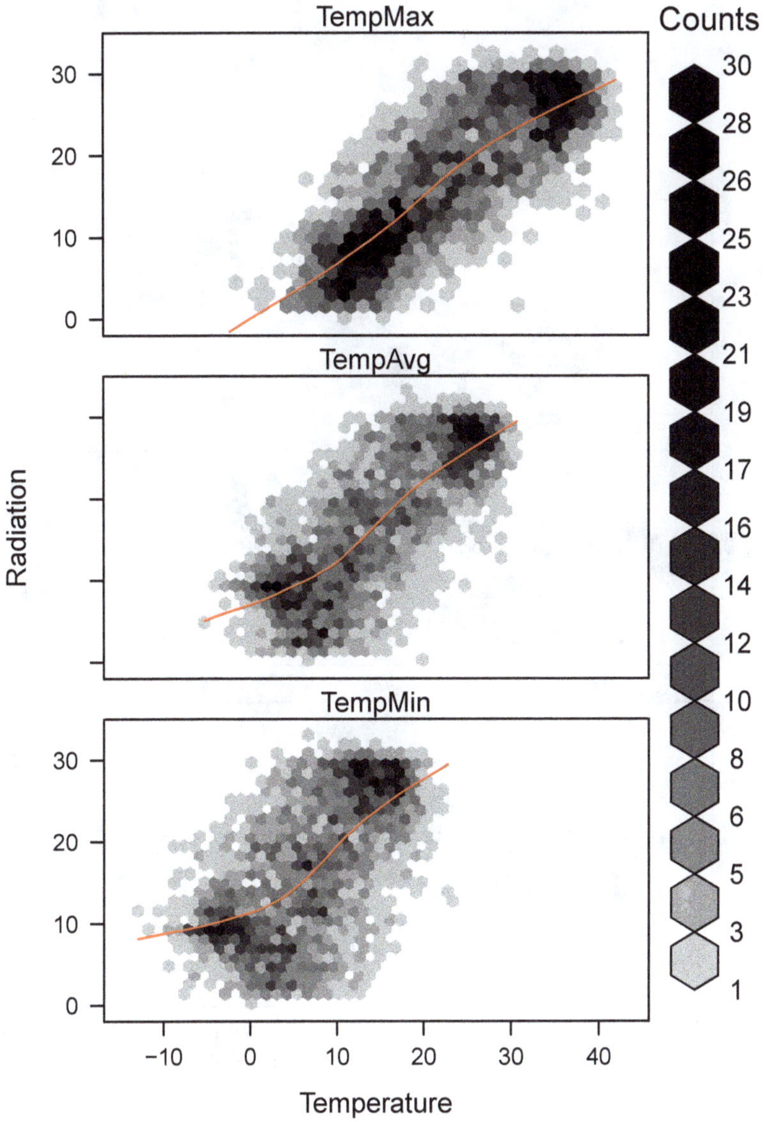

FIGURE 4.4: Scatterplot with hexagonal binning of temperature versus solar radiation using data of the Aranjuez station (lattice version).

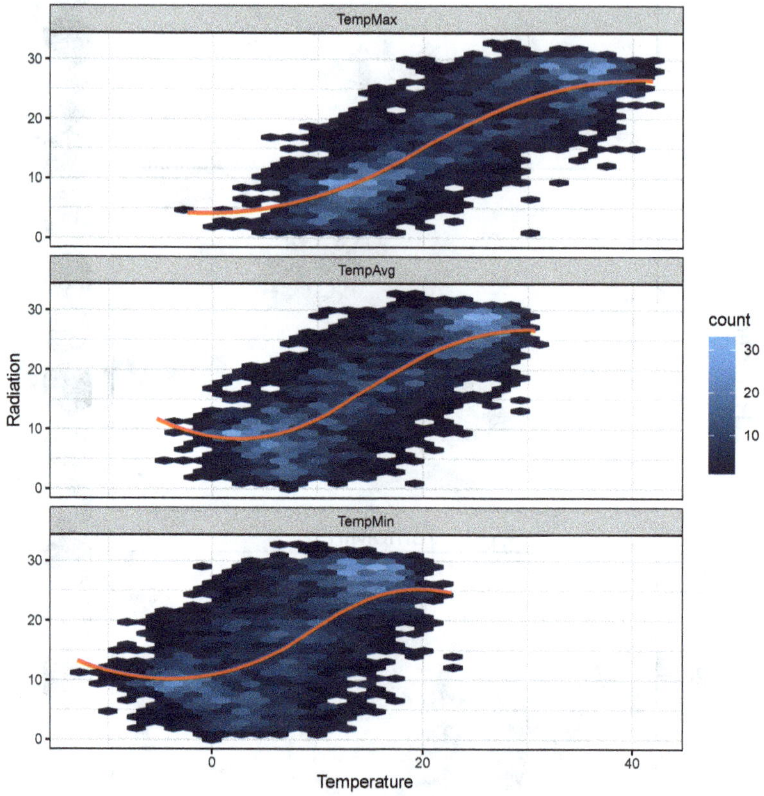

FIGURE 4.5: Scatterplot with hexagonal binning of temperature versus solar radiation using data of the Aranjuez station (ggplot2 version).

4.2 Scatterplot with Time as a Conditioning Variable

After discussing the hexagonal binning, let's recover the time variable. Figure 4.1 uses colors to encode months. Instead, we will now display separate scatterplots with a panel for each month. In addition, the statistic type (average, maximum, minimum) is included as an additional conditioning variable.

This matrix of panels can be displayed with ggplot using `facet_grid`. The code of Figure 4.6 uses partial transparency to cope with overplotting, small horizontal and vertical segments (`geom_rug`) to display point density on both variables, and a smooth line in each panel.

```
ggplot(data = aranjuezRshp, aes(Radiation, Temperature)) +
  facet_grid(Statistic ~ Month) +
  geom_point(col = 'skyblue4', pch = 19, cex = 0.5, alpha = 0.3) +
  geom_rug() +
  stat_smooth(se = FALSE, method = 'loess',
              col = 'indianred1', lwd = 1.2) +
  theme_bw() +
  theme(axis.text.x = element_text(size = 7, angle = 45))
```

The version with `lattice` needs the `useOuterStrips` function from the `latticeExtra` package, which prints the names of the conditioning variables on the top and left outer margins (Figure 4.7).

```
useOuterStrips(
  xyplot(Temperature ~ Radiation | Month * Statistic,
         data = aranjuezRshp,
         between = list(x = 0),
         col = 'skyblue4', pch = 19,
         cex = 0.5, alpha = 0.3)) +
  layer({
    panel.rug(..., col.line = 'indianred1',
              end = 0.05, alpha = 0.6)
    panel.loess(..., col = 'indianred1',
                lwd = 1.5, alpha = 1)
  })
```

These figures show the typical seasonal behavior of solar radiation and ambient temperature. Additionally, it displays in more detail the same relations between radiation and temperature already discussed with Figure 4.4.

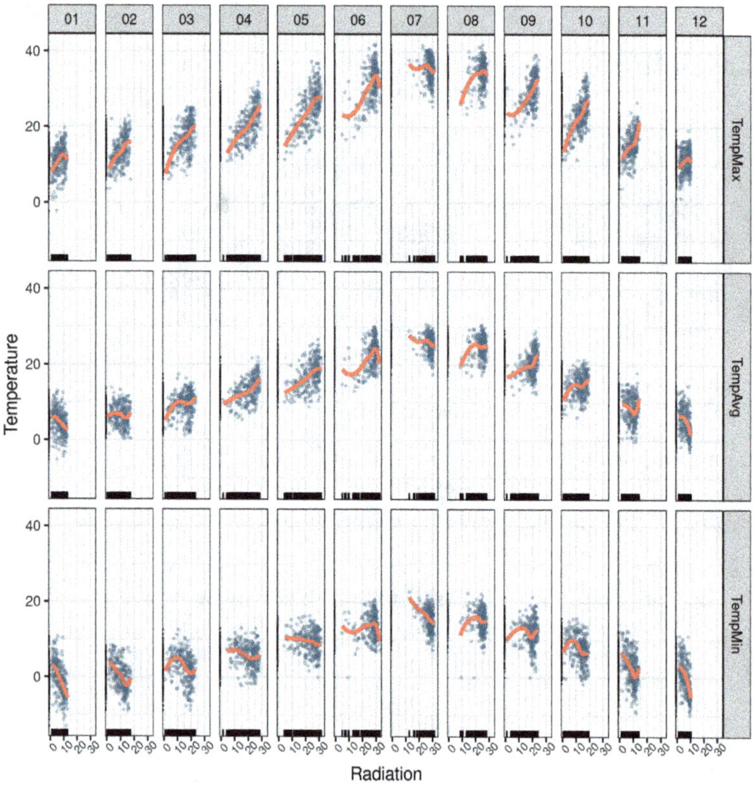

FIGURE 4.6: Scatterplot of temperature versus solar radiation for each month using data from the Aranjuez station (ggplot2 version).

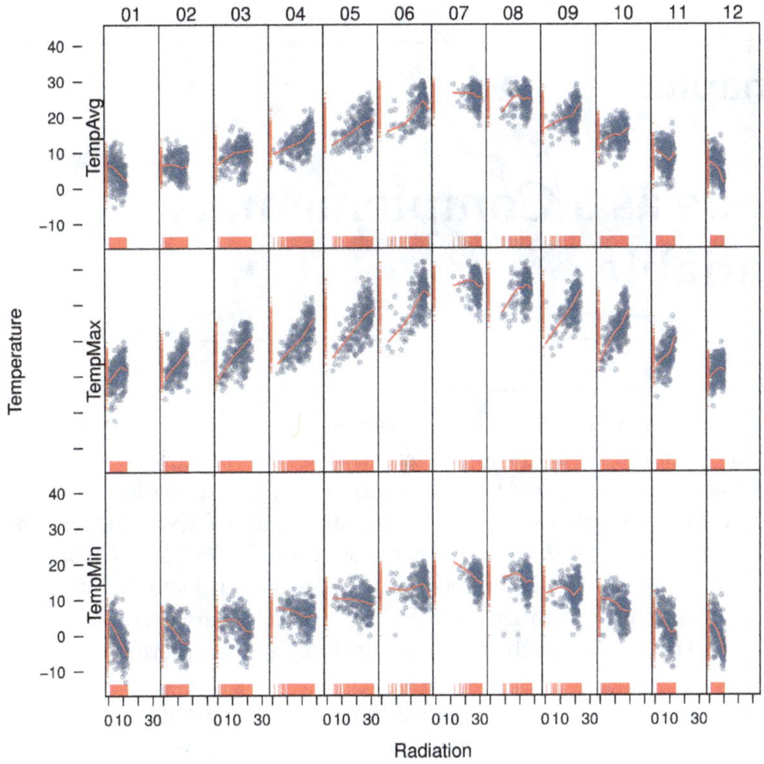

FIGURE 4.7: Scatterplot of temperature versus solar radiation for each month using data from the Aranjuez station (lattice version).

Chapter 5

Time as a Complementary Variable

In this chapter, time will be used as a complementary variable that adds information to a graph where several variables are confronted. We will illustrate this approach with the evolution of the relationship between Gross National Income (GNI) and carbon dioxide (CO_2) emissions for a set of countries extracted from the World Bank Open Data and the Climate Watch Data. We will try several solutions to display the relationship between CO_2 emissions and GNI over the years using time as a complementary variable.

Along this chapter, these subjects are covered: qualitative palettes, visual discrimination, panel functions, label positioning, small multiples, class intervals, and animation. The last section is devoted to interactive graphics.

The most relevant packages in this chapter are: zoo for reading and arranging data as time series; reshape2 for converting data from the wide format to the long format; RColorBrewer for defining color palettes; directlabels for label positioning; classInt for computing class intervals; plotly, gganimate, and gridSVG for interactive graphics.

DOI: 10.1201/9781003485384-5

5.1 Polylines

Our first approach is to display the entire data in a panel with a scatterplot using country names as the grouping factor. Points of each country are connected with polylines to reveal the time evolution (Figure 5.1).

```
library("zoo")

load('data/TimeSeries/CO2.RData')

## lattice version
xyplot(GNI.capita ~ CO2.capita, data = CO2data,
       xlab = "Carbon dioxide emissions (metric tons per capita)",
       ylab = "GNI per capita, PPP (current international $)",
       groups = Country.Name, type = 'b')

## ggplot2 version
ggplot(data = CO2data, aes(x = CO2.capita, y = GNI.capita,
                           color = Country.Name)) +
   xlab("Carbon dioxide emissions (metric tons per capita)") +
   ylab("GNI per capita, PPP (current international $)") +
   geom_point() + geom_path() + theme_bw()
```

Three improvements can be added to this graphical result:

1. Define a better palette to enhance visual discrimination between countries.

2. Display time information with labels to show year values.

3. Label each polyline with the country name instead of a legend.

5.1.1 Choosing Colors

The Country.Name categorical variable will be encoded with a qualitative palette, namely the first colors of the palette Set1[1] from the RColorBrewer package (Neuwirth 2022).

```
nCountries <- nlevels(CO2data$Country.Name)
pal <- brewer.pal(n = nCountries, 'Set1')
```

---

[1]http://colorbrewer2.org/

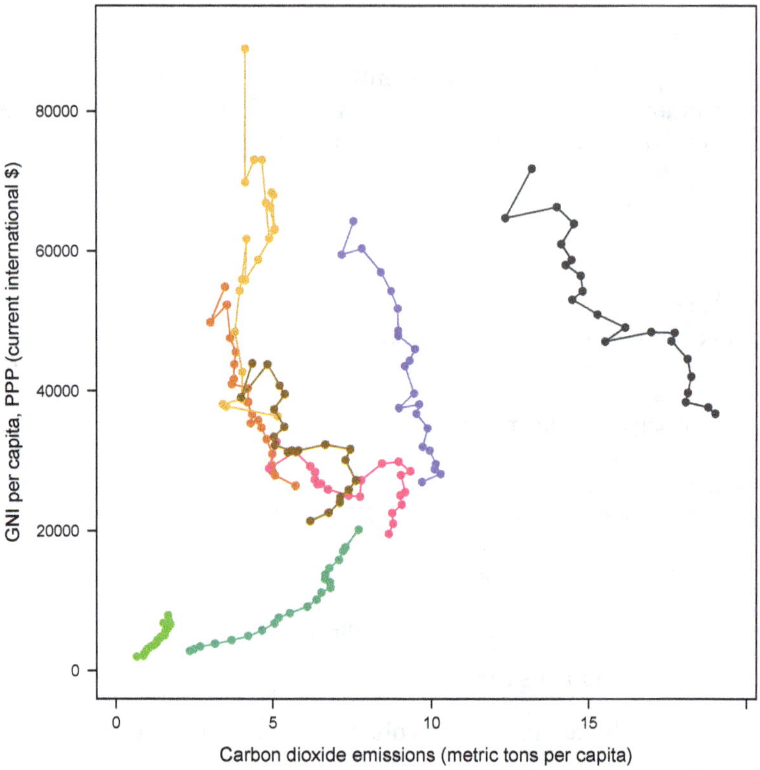

FIGURE 5.1: GNI per capita versus CO_2 emissions per capita.

Adjacent colors of this palette are chosen to be easily distinguishable. Therefore, the connection between colors and countries must be in such a way that nearby lines are encoded with adjacent colors of the palette.

A simple approach is to calculate the annual average of the variable to be represented along the x-axis (CO2.capita), and extract colors from the palette according to the order of this value.

```
## Rank of average values of CO2 per capita
CO2mean <- aggregate(CO2.capita ~ Country.Name,
                data = CO2data, FUN = mean)
palOrdered <- pal[rank(CO2mean$CO2.capita)]
```

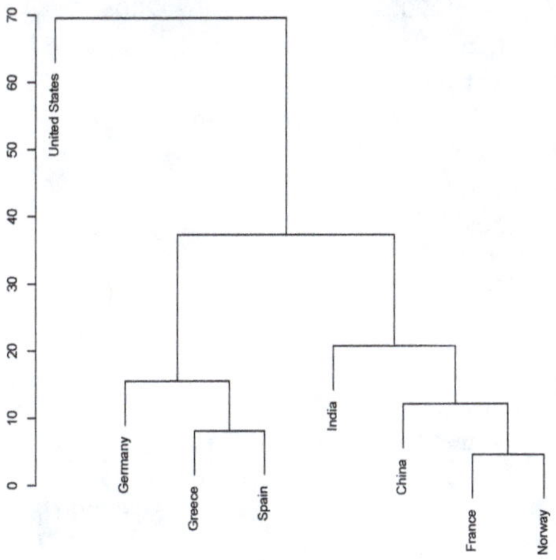

FIGURE 5.2: Hierarchical clustering of the time evolution of CO_2 per capita values.

A more sophisticated solution is to use the ordered results of a hierarchical clustering of the time evolution of the CO_2 per capita values (Figure 5.2). The data is extracted from the original CO_2 data.frame.

```
library("reshape2")

CO2capita <- CO2data[, c('Country.Name',
                         'Year',
                         'CO2.capita')]
CO2capita <- dcast(CO2capita, Country.Name ~ Year)

summary(CO2capita)
```

```
Using CO2.capita as value column: use value.var to override.
Country.Name        2000              2001              2002
  China  :1     Min.   : 0.6762   Min.   : 0.888    Min.   : 0.9011
  France :1     1st Qu.: 4.4675   1st Qu.: 3.265    1st Qu.: 3.2402
  Germany:1     Median : 5.9525   Median : 5.908    Median : 6.0131
  Greece :1     Mean   : 7.1857   Mean   : 7.076    Mean   : 6.9986
  India  :1     3rd Qu.: 8.9080   3rd Qu.: 9.161    3rd Qu.: 9.0882
  Norway :1     Max.   :19.0211   Max.   :18.797    Max.   :18.0839
  (Other):2
```

```
         2003                 2004                 2005
Min.    : 0.9092    Min.    : 0.9589    Min.    : 0.9873
1st Qu.: 3.8068    1st Qu.: 3.9477    1st Qu.: 4.1101
Median : 6.0474    Median : 6.1737    Median : 6.2871
Mean    : 7.1939    Mean    : 7.2800    Mean    : 7.3264
3rd Qu.: 9.3294    3rd Qu.: 9.2396    3rd Qu.: 9.3081
Max.    :18.1565    Max.    :18.2600    Max.    :18.1437


         2006                 2007                 2008
Min.    : 1.039    Min.    : 1.126    Min.    : 1.182
1st Qu.: 4.481    1st Qu.: 4.488    1st Qu.: 4.449
Median : 6.037    Median : 6.237    Median : 5.913
Mean    : 7.283    Mean    : 7.358    Mean    : 7.160
3rd Qu.: 9.242    3rd Qu.: 9.380    3rd Qu.: 9.119
Max.    :17.633    Max.    :17.742    Max.    :16.990


         2009                 2010                 2011                 2012
Min.    : 1.279    Min.    : 1.337    Min.    : 1.340    Min.    : 1.441
1st Qu.: 4.254    1st Qu.: 4.498    1st Qu.: 4.845    1st Qu.: 4.761
Median : 5.676    Median : 5.781    Median : 6.050    Median : 6.066
Mean    : 6.747    Mean    : 6.896    Mean    : 6.859    Mean    : 6.738
3rd Qu.: 8.558    3rd Qu.: 8.187    3rd Qu.: 8.087    3rd Qu.: 7.853
Max.    :15.531    Max.    :16.164    Max.    :15.291    Max.    :14.486


         2013                 2014                 2015
Min.    : 1.470    Min.    : 1.585    Min.    : 1.575
1st Qu.: 4.829    1st Qu.: 4.655    1st Qu.: 4.748
Median : 5.894    Median : 5.769    Median : 5.886
Mean    : 6.700    Mean    : 6.535    Mean    : 6.506
3rd Qu.: 7.472    3rd Qu.: 7.324    3rd Qu.: 7.210
Max.    :14.816    Max.    :14.755    Max.    :14.283


         2016                 2017                 2018
Min.    : 1.607    Min.    : 1.672    Min.    : 1.762
1st Qu.: 4.617    1st Qu.: 4.550    1st Qu.: 4.420
Median : 5.682    Median : 5.856    Median : 5.700
Mean    : 6.461    Mean    : 6.453    Mean    : 6.437
3rd Qu.: 7.218    3rd Qu.: 7.251    3rd Qu.: 7.398
Max.    :14.463    Max.    :14.140    Max.    :14.544


         2019                 2020                 2021
Min.    : 1.719    Min.    : 1.543    Min.    : 1.694
1st Qu.: 4.201    1st Qu.: 3.749    1st Qu.: 3.978
Median : 5.275    Median : 4.507    Median : 4.732
```

```
Mean    : 6.152    Mean    : 5.546    Mean    : 5.900
3rd Qu.: 7.342    3rd Qu.: 7.183    3rd Qu.: 7.565
Max.    :13.998    Max.    :12.363    Max.    :13.207
```

```r
hCO2 <- hclust(dist(CO2capita[, -1]))

oldpar <- par(mar = c(0, 2, 0, 0) + .1)
plot(hCO2, labels = CO2capita$Country.Name,
     xlab = '', ylab = '', sub = '', main = '')
par(oldpar)
```

The colors of the palette are assigned to each country with `match`, which returns a vector of the positions of the matches of the country names in alphabetical order in the country names ordered according to the hierarchical clustering.

```r
idx <- match(levels(CO2data$Country.Name),
             CO2capita$Country.Name[hCO2$order])
palOrdered <- pal[idx]
```

It must be highlighted that this palette links colors with the levels of `Country.Name` (country names in alphabetical order), which is exactly what the groups argument provides. The following code produces a curve for each country using different colors to distinguish them.

```r
## simpleTheme encapsulates the palette in a new theme for xyplot
myTheme <- simpleTheme(pch = 19, cex = 0.6, col = palOrdered)

## lattice version
pCO2.capita <- xyplot(GNI.capita ~ CO2.capita,
                   data = CO2data,
                   xlab = "Carbon dioxide emissions (metric tons
                       per capita)",
                   ylab = "GNI per capita, PPP (current
                       international $)",
                   groups = Country.Name,
                   par.settings = myTheme,
                   type = 'b')

## ggplot2 version
gCO2.capita <- ggplot(data = CO2data,
                   aes(x = CO2.capita,
                       y = GNI.capita,
                       color = Country.Name)) +
```

```
geom_point() + geom_path() +
scale_color_manual(values = palOrdered, guide = "none") +
xlab('CO2 emissions (metric tons per capita)') +
ylab('GNI per capita, PPP (current international $)') +
theme_bw()
```

5.1.2 Labels to Show Time Information

This result can be improved with labels displaying the years to show the time evolution. A panel function with `panel.text` to print the year labels and `panel.superpose` to display the lines for each group is a solution. In the panel function, `subscripts` is a vector with the integer indices representing the rows of the `data.frame` to be displayed in the panel.

```
xyplot(GNI.capita ~ CO2.capita,
    data = CO2data,
    xlab = "Carbon dioxide emissions (metric tons per capita)",
    ylab = "GNI per capita, PPP (current international $)",
    groups = Country.Name,
    par.settings = myTheme,
    type = 'b',
    panel = function(x, y, ..., subscripts, groups){
        panel.text(x, y, ...,
                   labels = CO2data$Year[subscripts],
                   pos = 2, cex = 0.5, col = 'gray')
        panel.superpose(x, y, subscripts, groups,...)
    })
```

The same result with a clearer code is obtained with the combination of `+.trellis`, `glayer_` and `panel.text`. Using `glayer_` instead of `glayer`, we ensure that the labels are printed below the lines.

```
## lattice version
pCO2.capita <- pCO2.capita +
    glayer_(panel.text(...,
                labels = CO2data$Year[subscripts],
                pos = 2, cex = 0.5, col = 'gray'))
```

```
## ggplot2 version
gCO2.capita <- gCO2.capita + geom_text(aes(label = Year),
                                colour = 'gray',
                                size = 2.5,
                                hjust = 0, vjust = 0)
```

5.1.3 Country Names: Positioning Labels

The common solution to link each curve with the group value is to add a legend. However, a legend can be confusing with too many items. In addition, the reader must carry out a complex task: Choose the line, memorize its color, search for it in the legend, and read the country name.

A better approach is to label each line using nearby text with the same color encoding. A suitable method is to place the labels avoiding the overlapping between labels and lines. The package directlabels (Hocking 2024) includes a wide repertory of positioning methods to cope with overlapping. The main function, direct.label, is able to determine a suitable method for each plot, although the user can choose a different method from the collection or even define a custom method. For the pCO2.capita object, the best results are obtained with extreme.grid (Figure 5.3).

```
library("directlabels")

## lattice version
direct.label(pCO2.capita,
            method = 'extreme.grid')

## ggplot2 version
direct.label(gCO2.capita, method = 'extreme.grid')
```

5.2 A Panel for Each Year

Time can be used as a conditioning variable (as shown in previous sections) to display subsets of the data in different panels. Figure 5.4 is produced with the same code as in Figure 5.1, now including |factor(Year) in the lattice version and facet_wrap(~ Year) in the ggplot2 version.

```
## lattice version
xyplot(GNI.capita ~ CO2.capita | factor(Year),
    data = CO2data,
    xlab = "Carbon dioxide emissions (metric tons per capita)",
    ylab = "GNI per capita, PPP (current international $)",
    groups = Country.Name, type = 'p',
    auto.key = list(space = 'right'))

## ggplot2 version
ggplot(data = CO2data,
    aes(x = CO2.capita,
```

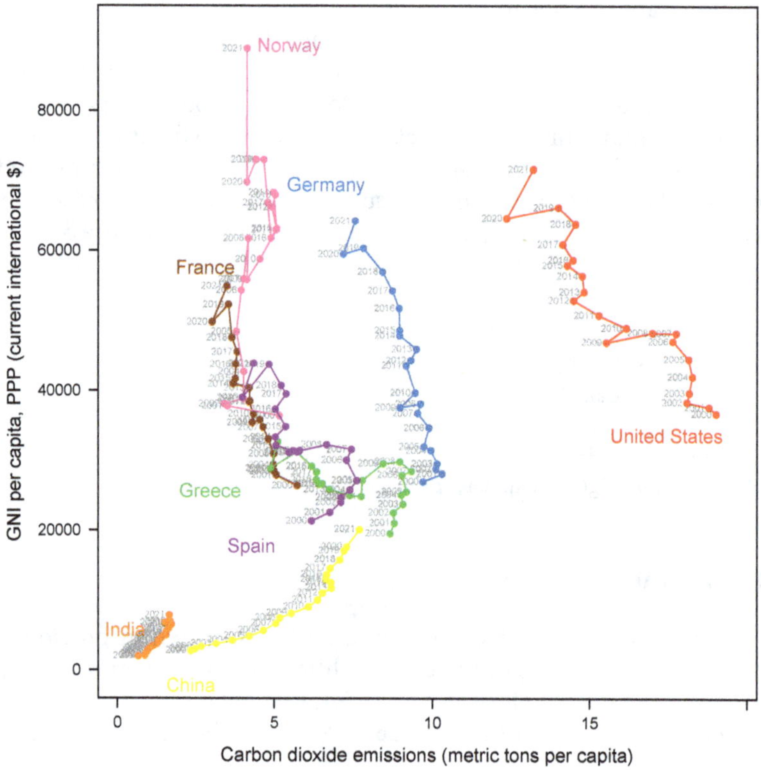

FIGURE 5.3: CO$_2$ emissions versus GNI per capita. Labels are placed with the `extreme.grid` method of the `directlabels` package.

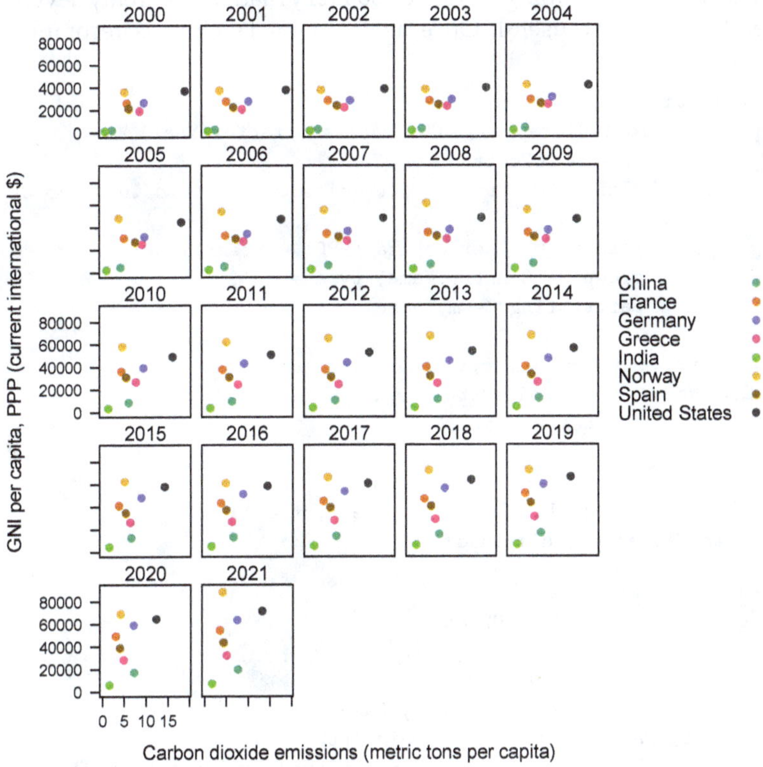

FIGURE 5.4: CO$_2$ emissions versus GNI per capita with a panel for each year.

```
              y = GNI.capita,
              colour = Country.Name)) +
      facet_wrap(~ Year) + geom_point(pch = 19) +
      xlab('CO2 emissions (metric tons per capita)') +
      ylab('GNI per capita, PPP (current international $)') +
      theme_bw()
```

Because the grouping variable, `Country.Name`, has many levels, the legend is not very useful. Once again, point labeling is recommended (Figure 5.5).

```
## lattice version
pl <- xyplot(GNI.capita ~ CO2.capita | factor(Year),
             data = CO2data,
             xlab = "Carbon dioxide emissions (metric tons per capita
                 )",
             ylab = "GNI per capita, PPP (current international $)",
             groups = Country.Name, type = 'b',
             par.settings = myTheme)

## Extracted from the qp.labels help page
dlmethod <- list("first.points",
                 cex = 0.5,
                 "calc.boxes",
                 "enlarge.box",
                 qp.labels("y","bottom","top"))
direct.label(pl, dlmethod)

## ggplot2 version
pg <- ggplot(data = CO2data,
             aes(x = CO2.capita,
                 y = GNI.capita,
                 colour = Country.Name)) +
     facet_wrap(~ Year) + geom_point(pch = 19) +
     scale_color_manual(values = palOrdered, guide = "none") +
     xlab('CO2 emissions (metric tons per capita)') +
     ylab('GNI per capita, PPP (current international $)') +
     theme_bw()

direct.label(pg, dlmethod)
```

5.2.1 ✍Using Variable Size to Encode an Additional Variable

Instead of using simple points, we can display circles of different radius to encode a new variable. This new variable is `CO2.PPP`, the ratio of CO_2

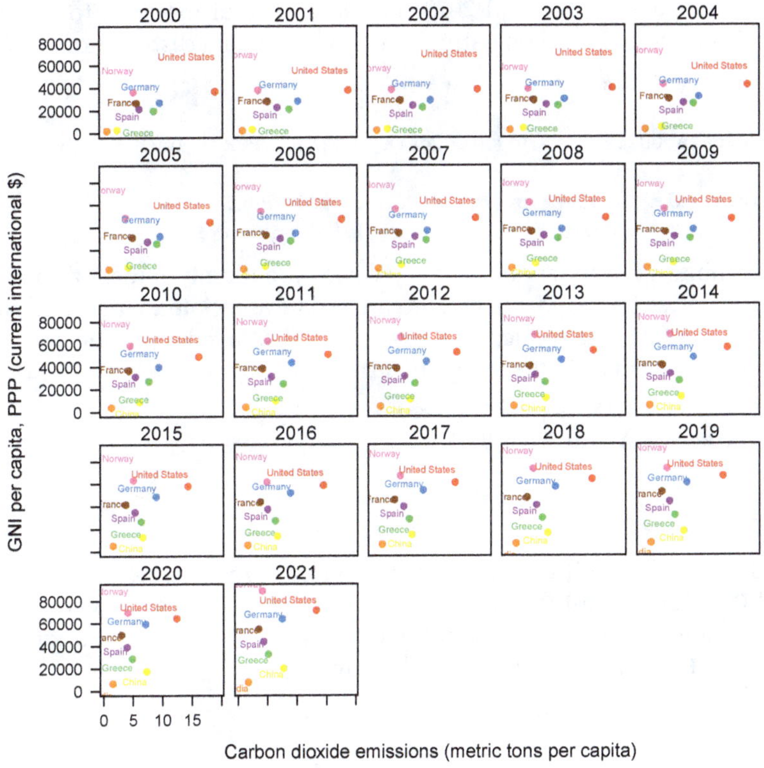

FIGURE 5.5: CO_2 emissions versus GNI per capita with a panel for each year.

emissions to the Gross Domestic Product with purchasing power parity (PPP) estimations.

To use this numeric variable as an additional grouping factor, its range must be divided into different classes. The typical solution is to use cut to coerce the numeric variable into a factor whose levels correspond to uniform intervals, which could be unrelated to the data distribution. The classInt package (Bivand 2023) provides several methods to partition data into classes based on natural groups in the data distribution.

```
library("classInt")

CO2data$CO2.PPP <- with(CO2data, CO2/GNI.PPP * 1e9)

intervals <- classIntervals(CO2data$CO2.PPP,
                            n = 4, style = 'fisher')
```

Although the functions of this package are mainly intended to create color palettes for maps, the results can also be associated with point sizes. cex.key defines the sequence of sizes (to be displayed in the legend) associated with each CO2.PPP using the findCols function.

```
nInt <- length(intervals$brks) - 1
cex.key <- seq(0.5, 1.8, length = nInt)

idx <- findCols(intervals)
CO2data$cexPoints <- cex.key[idx]
```

The graphic will display information on two variables (GNI.capita and CO2.capita in the vertical and horizontal axes, respectively) with a conditioning variable (Year) and two grouping variables (Country.Name, and CO2.PPP through cexPoints) (Figure 5.6).

```
ggplot(data = CO2data,
       aes(x = CO2.capita,
           y = GNI.capita,
           colour = Country.Name)) +
   facet_wrap(~ Year) +
   geom_point(aes(size = cexPoints), pch = 19) +
   xlab('Carbon dioxide emissions (metric tons per capita)') +
   ylab('GNI per capita, PPP (current international $)') +
   theme_bw()
```

The auto.key mechanism of the lattice version is not able to cope with two grouping variables. Therefore, the legend, whose main componens are the labels (intervals) and the point sizes (cex.key), must be defined manually (Figure 5.7).

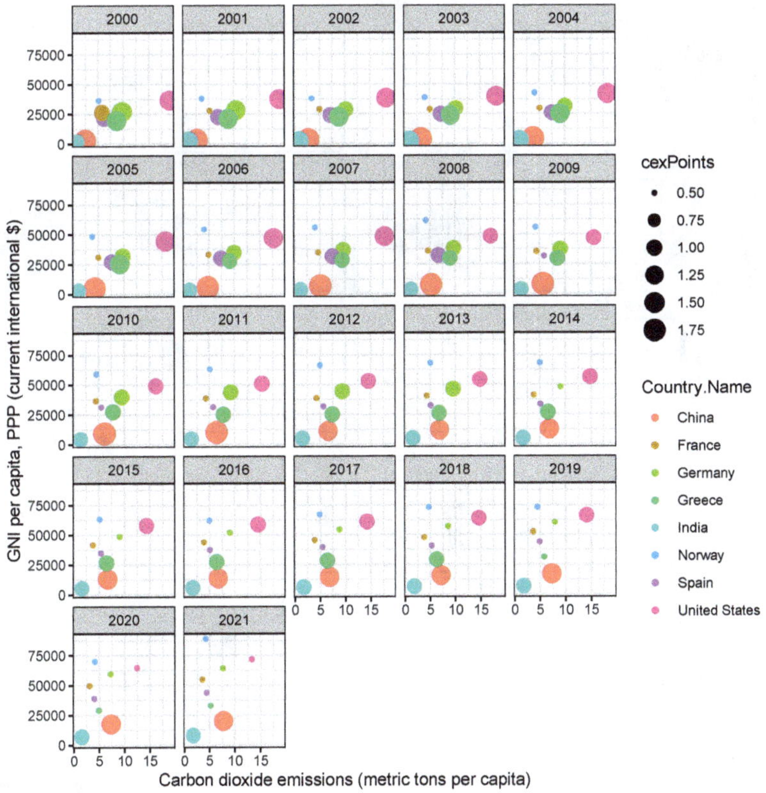

FIGURE 5.6: CO_2 emissions versus GNI per capita for different intervals of the ratio of CO_2 emissions to the GDP PPP estimations.

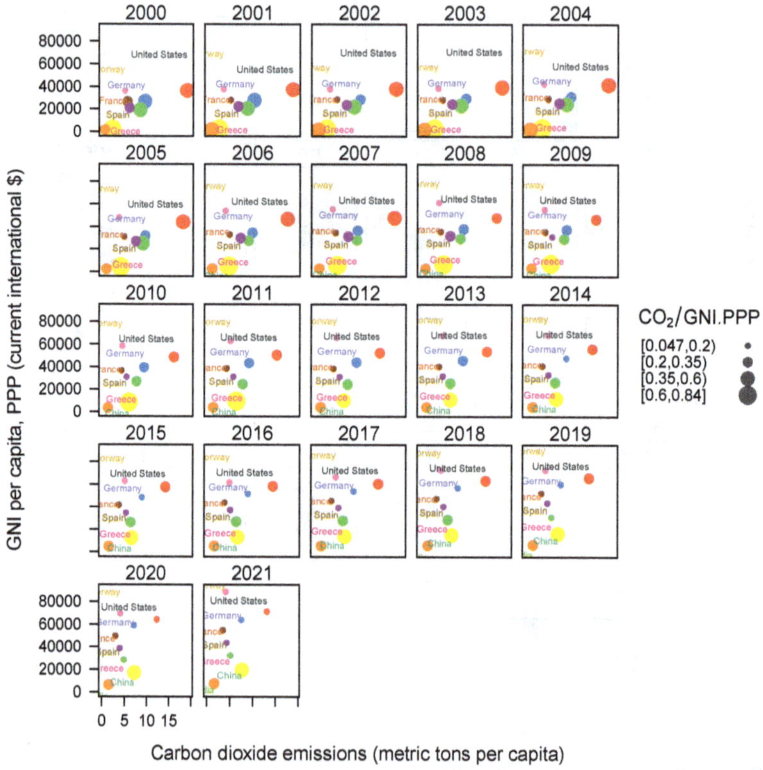

FIGURE 5.7: CO$_2$ emissions versus GNI per capita for different intervals of the ratio of CO$_2$ emissions to the GDP PPP estimations.

```r
op <- options(digits = 2)
tab <- print(intervals)
options(op)

key <- list(space = 'right',
        title = expression(CO[2]/GNI.PPP),
        cex.title = 1,
        ## Labels of the key are the intervals strings
        text = list(labels = names(tab), cex = 0.85),
        ## Points sizes are defined with cex.key
        points = list(col = 'black',
                    pch = 19,
                    cex = cex.key,
                    alpha = 0.7)
        )

pl2 <- xyplot(GNI.capita ~ CO2.capita|factor(Year), data =
    CO2data,
        xlab = "Carbon dioxide emissions (metric tons per
            capita)",
        ylab = "GNI per capita, PPP (current international $)"
            ,
        groups = Country.Name,
        alpha = 0.7,
        panel = panel.superpose,
        panel.groups = function(x, y,
                            subscripts, group.number,
                                group.value, ...){
            panel.xyplot(x, y,
                    col = palOrdered[group.number],
                    cex = CO2data$cexPoints[subscripts])
            })
## Add labels...
plLabel <- direct.label(pl2, "dlmethod")
## ... and the key (after the call to direct.labels because this
## function removes the legend)
update(plLabel, key = key)
```

5.3   Interactive Graphics: Animation

Previous sections have been focused on static graphics. This section describes several solutions to display the data through animation with interactive functionalities.

Gapminder[2] is an independent foundation based in Stockholm, Sweden. Its mission is "to debunk devastating myths about the world by offering free access to a fact-based world view." They provide free online tools, data, and videos "to better understand the changing world." The initial development of Gapminder was the Trendalyzer software, used by Hans Rosling in several sequences of his documentary "The Joy of Stats."

The information visualization technique used by Trendalyzer is an interactive bubble chart. By default it shows five variables: two numeric variables on the vertical and horizontal axes, bubble size and color, and a time variable that may be manipulated with a slider. The software uses brushing and linking techniques for displaying the numeric value of a highlighted country.

We will mimic the Trendalyzer/Motion Chart solution, using traveling bubbles of different colors and with radii proportional to the values of the variable CO2.PPP. Previously, you should watch the magistral video "200 Countries, 200 Years, 4 Minutes"[3].

Three packages are used here: plotly, gganimate, and gridSVG.

5.3.1 plotly

The plotly package has already been used in Section 3.4.3 to create an interactive graphic representing time on the x-axis. In this section, this package produces an animation piping the result of the plot_ly and add_markers functions through the animation_slider function.

Variables CO2.capita and GNI.capita are represented in the x-axis and y-axis, respectively.

```
library("plotly")

p <- plot_ly(CO2data,
             x = ~CO2.capita,
             y = ~GNI.capita,
             sizes = c(10, 100),
             marker = list(opacity = 0.7,
                           sizemode = 'diameter'))
```

CO2.PPP is encoded with the circle sizes, while Country.Name is represented with colors and labels.

---

[2]https://www.gapminder.org/
[3]https://www.gapminder.org/videos/200-years-that-changed-the-world/

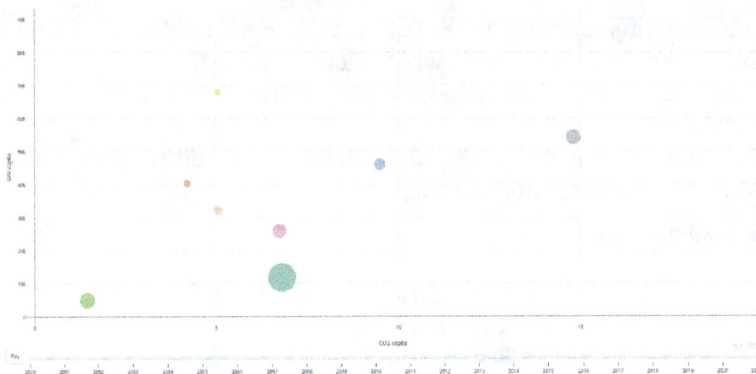

FIGURE 5.8: Snapshot of a Motion Chart produced with `plotly`.

```
p <- add_markers(p,
            size = ~CO2.PPP,
            color = ~Country.Name,
            text = ~Country.Name, hoverinfo = "text",
            ids = ~Country.Name,
            frame = ~Year,
            showlegend = FALSE)
```

Finally, animation is created with `animation_opts`, to customize the frame and transition times and with `animation_slider` to define the slider. Figure 5.8 is a snapshot of this animation.

```
p <- animation_opts(p,
                frame = 1000,
                transition = 800,
                redraw = FALSE)

p <- animation_slider(p,
                currentvalue = list(prefix = "Year "))


p
```

5.3.2 gganimate

The gganimate package expands the ecosystem of `ggplot2`, focused on static graphics, with new grammar classes for animation. In this example we will use three of them: `transition_time`, which defines how the data evolves over time; `ease_aes`, which defines how the aesthetics change with transitions; and `shadow_wake`, which defines how to render data from past instants.

The animation is initiated with a conventional call to `ggplot` and the corresponding `geom_*` and `scale_*` functions.

```r
library("gganimate")

pgvis <- ggplot(data = CO2data,
                aes(x = CO2.capita,
                    y = GNI.capita,
                    size = CO2.PPP,
                    color = Country.Name)) +
    geom_point(pch = 19) +
    scale_size(range = c(1, 12)) +
    scale_color_viridis_d() +
    theme_bw() +
    labs(title = 'Year: {trunc(frame_time)}',
        x = 'CO2 emissions (metric tons per capita)',
        y = 'GNI per capita, PPP (current international $)')
```

This call produces a static graphic that is transformed in an animation with the adition of `transition_time` based on the `Year` variable. The animation contained in the resulting gganim object can be rendered with the `animate` function. Figure 5.9 displays a snapshot of this animation.

```r
pgAnim <- pgvis +
    transition_time(Year) +
    ease_aes("linear") +
    shadow_wake(wake_length = 0.3)

animate(pgAnim,
        height = 1080, width = 1080,
        res = 150,
        units = "px")

anim_save("figs/TimeSeries/CO2_gganimate.gif")
```

5.3.3 gridSVG

The final solution to create an animation is based on the function `grid.animate` of the `gridSVG` package.

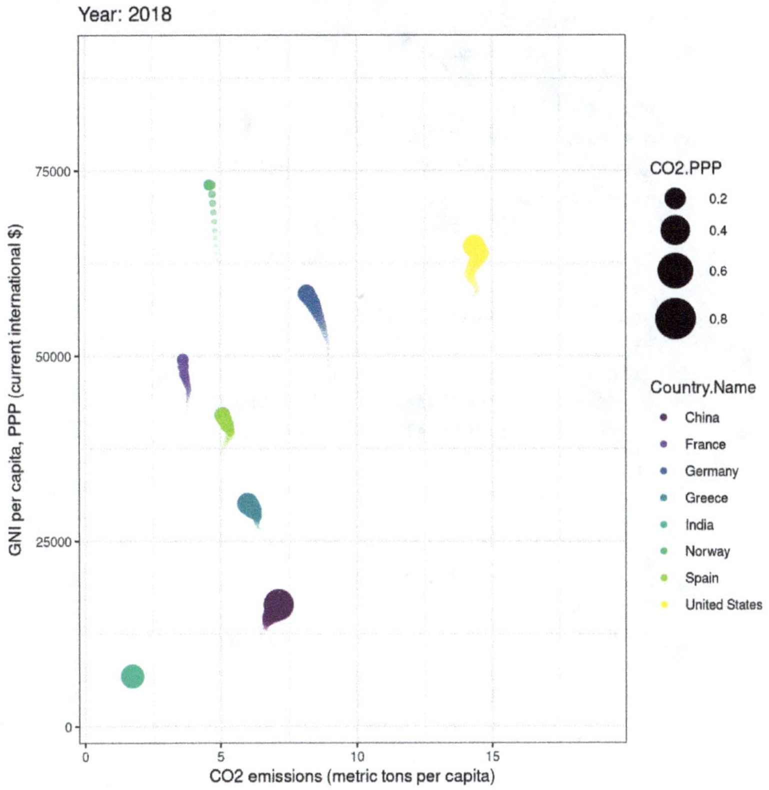

FIGURE 5.9: Snapshot of a motion chart produced with gganimate.

The first step is to draw the initial state of the bubbles. Their colors are again defined by the pal0rdered palette (line 15), although the adjust-color function is used for a lighter fill color. Because there will not be a legend, there is no need to define class intervals, and thus the radius is directly proportional to the value of CO2data$CO2.PPP (line 16).

```
1  library("gridSVG")
2  library("grid")
3
4  xyplot(GNI.capita ~ CO2.capita,
5        data = CO2data,
6        xlab = "Carbon dioxide emissions (metric tons per capita)",
7        ylab = "GNI per capita, PPP (current international $)",
```

```
8       subset = Year==2000,
9       groups = Country.Name,
10      ## The limits of the graphic are defined
11      ## with the entire dataset
12      xlim = extendrange(CO2data$CO2.capita),
13      ylim = extendrange(CO2data$GNI.capita),
14      panel = function(x, y, ..., subscripts, groups) {
15          color <- palOrdered[groups[subscripts]]
16          radius <- CO2data$CO2.PPP[subscripts]
17          ## Size of labels
18          cex <- 1.1*sqrt(radius)
19          ## Bubbles
20          grid.circle(x, y, default.units = "native",
21                      r = radius*unit(.25, "inch"),
22                      name = trellis.grobname("points", type = "panel
                            "),
23                      gp = gpar(col = color,
24                              ## Fill color ligther than border
25                              fill = adjustcolor(color, alpha = .5),
26                              lwd = 2))
27          ## Country labels
28          grid.text(label = groups[subscripts],
29                    x = unit(x, 'native'),
30                    ## Labels above each bubble
31                    y = unit(y, 'native') + 1.5 * radius *unit(.25,
                            'inch'),
32                    name = trellis.grobname('labels', type = 'panel'
                            ),
33                    gp = gpar(col = color, cex = cex))
34      })
```

From this initial state, `grid.animate` creates a collection of animated graphical objects with the result of `animUnit` (lines 9, 11, 14, and 18). This function produces a set of values that will be interpreted by `grid.animate` as intermediate states of a feature of the graphical object (lines 21 and 29). Thus, the bubbles will travel across the values defined by `x_points` and `y_points`, while their labels will use `x_points` and `x_labels`.

The use of `rep=TRUE` ensures that the animation will be repeated indefinitely (lines 27 and 34).

```
1   ## Duration in seconds of the animation
2   duration <- 20
3
4   nCountries <- nlevels(CO2data$Country.Name)
```

```
 5  years <- unique(CO2data$Year)
 6  nYears <- length(years)
 7
 8  ## Intermediate positions of the bubbles
 9  x_points <- animUnit(unit(CO2data$CO2.capita, 'native'),
10                id = rep(seq_len(nCountries), each = nYears))
11  y_points <- animUnit(unit(CO2data$GNI.capita, 'native'),
12                id = rep(seq_len(nCountries), each = nYears))
13  ## Intermediate positions of the labels
14  y_labels <- animUnit(unit(CO2data$GNI.capita, 'native') +
15                1.5 * CO2data$CO2.PPP * unit(.25, 'inch'),
16                id = rep(seq_len(nCountries), each = nYears))
17  ## Intermediate sizes of the bubbles
18  size <- animUnit(CO2data$CO2.PPP * unit(.25, 'inch'),
19                id = rep(seq_len(nCountries), each = nYears))
20
21  grid.animate(trellis.grobname("points", type = "panel",
22                      row = 1, col = 1),
23            duration = duration,
24            x = x_points,
25            y = y_points,
26            r = size,
27            rep = TRUE)
28
29  grid.animate(trellis.grobname("labels", type = "panel",
30                      row = 1, col = 1),
31            duration = duration,
32            x = x_points,
33            y = y_labels,
34            rep = TRUE)
```

A bit of interactivity can be added with the grid.hyperlink function. For example, the following code adds the corresponding Wikipedia link to a mouse click on each bubble.

```
countries <- unique(CO2data$Country.Name)
URL <- paste('http://en.wikipedia.org/wiki/', countries, sep = ''
    )
grid.hyperlink(trellis.grobname('points', type = 'panel', row =
    1, col = 1),
          URL, group = FALSE)
```

Finally, the time information: The year is printed in the lower right corner, using the visibility attribute of an animated textGrob object to show and hide the values.

```
visibility <- matrix("hidden", nrow = nYears, ncol = nYears)
diag(visibility) <- "visible"
yearText <- animateGrob(garnishGrob(textGrob(years, .9, .15,
                                    name = "year",
                                    gp = gpar(cex = 2, col = "
                                        grey")),
                        visibility = "hidden"),
                duration = 20,
                visibility = visibility,
                rep = TRUE)
grid.draw(yearText)
```

The SVG file produced with grid.export is available at the website of the book (Figure 5.10). Because this animation does not trace the paths, Figure 5.3 provides this information as a static complement.

```
grid.export("figs/TimeSeries/bubbles.svg")
```

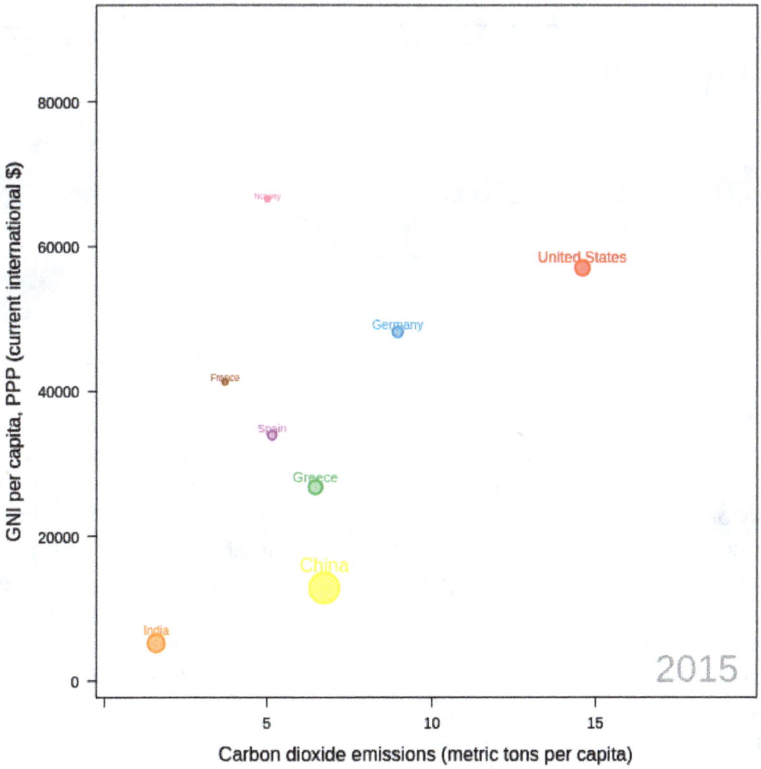

FIGURE 5.10: Animated bubbles produced with gridSVG.

Chapter 6

About the Data

6.1 SIAR

The Agroclimatic Information System for Irrigation (SIAR) (MAGRAMA 2024) is a free-download database operating since 1999, covering the majority of the irrigated area of Spain. This network belongs to the Ministry of Agriculture, Food, and Environment of Spain as a tool to predict and study meteorological variables for agriculture. SIAR is composed of twelve regional centers and a national center, aiming to centralize and depurate measurements from the stations of the network. Figure 6.1 displays the stations over an altitude map. Some stations from the complete network have been omitted due to difficulties accessing their coordinates or to incomplete or spurious data series[1].

6.1.1 Daily Data of Different Meteorological Variables

As an example of multiple time series with different scales, we will use eight years (from January 2004 to December 2011) of daily data corresponding to several meteorological variables measured at the SIAR station located at Aranjuez (Madrid, Spain), available on the SIAR webpage[2].

---

[1]The name and location data of these stations are available at the GitHub repository of the paper (Antonanzas-Torres, Cañizares, and O. Perpiñán 2013).

[2]https://servicio.mapa.gob.es/websiar/

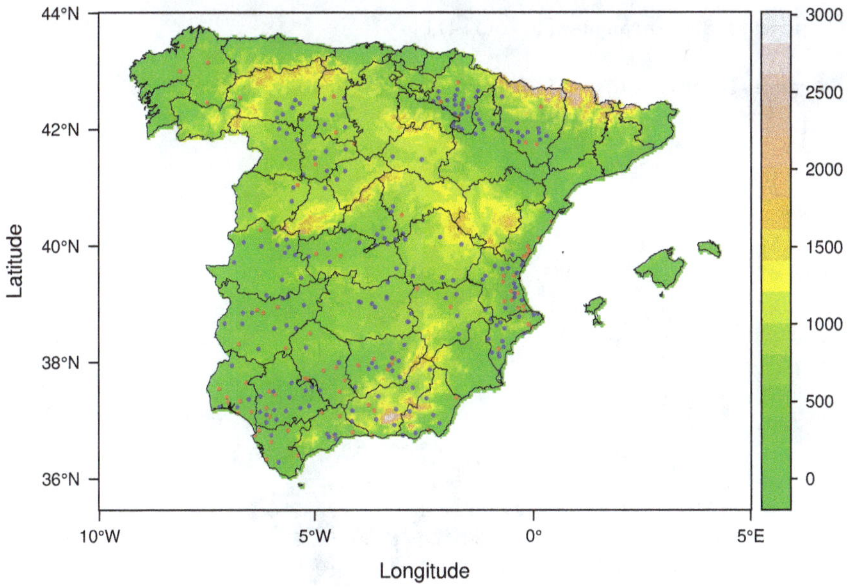

FIGURE 6.1: Meteorological stations of the SIAR network. The color key indicates the altitude (meters).

The `aranjuez.gz` file, available in the `data` folder of the book repository, contains this information with several meteorological variables: average, maximum, and minimum ambient temperature; average and maximum humidity; average and maximum wind speed; rainfall; solar radiation on the horizontal plane; and evapotranspiration.

The `read.zoo` from the `zoo` package accepts this string and downloads the data to construct a `zoo` object. Several arguments are passed directly to `read.table` (`header`, `skip`, etc.) and are detailed conveniently on the help page of this function. The `index.column` is the number of the column with the time index, and `format` defines the date format of this index.

```
library("zoo")

aranjuez <- read.zoo("data/TimeSeries/aranjuez.gz",
                index.column = 3, format = "%d/%m/%Y",
                fileEncoding = 'UTF-16LE',
                header = TRUE, fill = TRUE,
                sep = ';', dec = ",", as.is = TRUE)
```

91

```
aranjuez <- aranjuez[, -c(1:4)]

names(aranjuez) <- c('TempAvg', 'TempMax', 'TempMin',
                     'HumidAvg', 'HumidMax',
                     'WindAvg', 'WindMax',
                     'Radiation', 'Rain', 'ET')
```

```
     Index                 TempAvg              TempMax              TempMin
Min.   :2004-01-01   Min.    :-5.310   Min.    :-2.36    Min.    :-12.980
1st Qu.:2005-12-29   1st Qu.: 7.692    1st Qu.:14.53     1st Qu.:  1.520
Median :2008-01-09   Median :13.810    Median :21.67     Median :  7.170
Mean   :2008-01-03   Mean    :14.405   Mean    :22.54    Mean    :  6.894
3rd Qu.:2010-01-02   3rd Qu.:21.615    3rd Qu.:30.89     3rd Qu.: 12.590
Max.   :2011-12-31   Max.    :30.680   Max.    :41.91    Max.    : 22.710
                                       NA's    :3        NA's    :7
    HumidAvg            HumidMax             WindAvg              WindMax
Min.   :19.89     Min.    : 35.88   Min.    :0.250    Min.    : 1.550
1st Qu.:47.04     1st Qu.: 81.60    1st Qu.:0.670     1st Qu.: 3.840
Median :62.49     Median : 90.90    Median :0.920     Median : 5.150
Mean   :62.11     Mean    : 87.21   Mean    :1.166    Mean    : 5.467
3rd Qu.:77.30     3rd Qu.: 94.90    3rd Qu.:1.430     3rd Qu.: 6.760
Max.   :99.50     Max.    :103.00   Max.    :6.450    Max.    :18.060
NA's   :2         NA's    :10
   Radiation          Rain                 ET
Min.   : 0.28     Min.    : 0.000   Min.    :0.000
1st Qu.: 9.37     1st Qu.: 0.000    1st Qu.:1.160
Median :16.67     Median : 0.000    Median :2.750
Mean   :16.73     Mean    : 1.046   Mean    :3.088
3rd Qu.:24.63     3rd Qu.: 0.200    3rd Qu.:4.923
Max.   :32.74     Max.    :49.730   Max.    :8.560
                                    NA's    :14
```

From the summary it is clear that parts of these time series include erroneous outliers that can be safely removed:

```
aranjuezClean <- within(as.data.frame(aranjuez),{
   TempMin[TempMin > 40] <- NA
   HumidMax[HumidMax > 100] <- NA
})

aranjuez <- zoo(aranjuezClean, index(aranjuez))
```

```
     Index              TempAvg            TempMax            TempMin
Min.    :2004-01-01  Min.    :-5.310  Min.    :-2.36   Min.    :-12.980
1st Qu.:2005-12-29  1st Qu.: 7.692  1st Qu.:14.53   1st Qu.:  1.520
Median :2008-01-09  Median :13.810  Median :21.67   Median :  7.170
Mean   :2008-01-03  Mean   :14.405  Mean   :22.54   Mean   :  6.894
3rd Qu.:2010-01-02  3rd Qu.:21.615  3rd Qu.:30.89   3rd Qu.: 12.590
Max.    :2011-12-31  Max.    :30.680  Max.    :41.91   Max.    : 22.710
                                     NA's    :3       NA's    :7

   HumidAvg           HumidMax           WindAvg            WindMax
Min.    :19.89    Min.    : 35.88   Min.    :0.250   Min.    : 1.550
1st Qu.:47.04    1st Qu.: 81.60   1st Qu.:0.670   1st Qu.: 3.840
Median :62.49    Median : 90.90   Median :0.920   Median : 5.150
Mean   :62.11    Mean   : 87.20   Mean   :1.166   Mean   : 5.467
3rd Qu.:77.30    3rd Qu.: 94.90   3rd Qu.:1.430   3rd Qu.: 6.760
Max.    :99.50    Max.    :100.00   Max.    :6.450   Max.    :18.060
NA's    :2       NA's    :11
   Radiation          Rain               ET
Min.    : 0.28    Min.    : 0.000   Min.    :0.000
1st Qu.: 9.37    1st Qu.: 0.000   1st Qu.:1.160
Median :16.67    Median : 0.000   Median :2.750
Mean   :16.73    Mean   : 1.046   Mean   :3.088
3rd Qu.:24.63    3rd Qu.: 0.200   3rd Qu.:4.923
Max.    :32.74    Max.    :49.730   Max.    :8.560
                                     NA's    :14
```

6.1.2 Solar Radiation Measurements from Different Locations

As an example of multiple time series with the same scale, we will use data of daily solar radiation measurements from different locations.

Daily solar radiation incident on the horizontal plane is registered by meteorological stations and estimated from satellite images. This meteorological variable is important for a wide variety of scientific disciplines and engineering applications. Its variations and trends, dependent on the location (mainly latitude, and also longitude and altitude) and on time (day of the year), have been analyzed and modeled in a huge collection of papers and reports. In this section we will focus our attention on the time evolution of the solar radiation. The spatial distribution and the spatio-temporal behavior will be the subject of later sections.

The stations of the SIAR network include first-class pyranometers according to the World Meteorological Organization (WMO), whose absolute accuracy is within ±5% and is typically lower than ±3%. Solar irradiance is recorded every 15 minutes and then collated through a datalogger within the station to generate the daily irradiation, which is later sent to the regional and national centers.

The file navarra.RData contains daily solar radiation data of 2011 from the meteorological stations of Navarra, Spain. The names of the dataset are the abbreviations of each station name.

```
library("zoo")

load('data/TimeSeries/navarra.RData')
```

```
      Index                Arzr                 Adó                 Lmbr
Min.   :2011-01-01   Min.   : 1.562   Min.   : 0.028   Min.   : 2.122
1st Qu.:2011-04-02   1st Qu.: 8.680   1st Qu.: 7.630   1st Qu.: 8.610
Median :2011-07-02   Median :16.770   Median :15.680   Median :17.080
Mean   :2011-07-02   Mean   :16.627   Mean   :15.717   Mean   :16.767
3rd Qu.:2011-10-01   3rd Qu.:24.590   3rd Qu.:23.970   3rd Qu.:24.800
Max.   :2011-12-31   Max.   :32.400   Max.   :32.060   Max.   :32.820

      Ancn                 Artj                Aibr                 SMdU
Min.   : 0.009   Min.   : 1.32   Min.   : 0.94   Min.   : 0.971
1st Qu.: 6.387   1st Qu.: 8.26   1st Qu.: 7.99   1st Qu.: 8.075
Median :13.950   Median :15.43   Median :16.08   Median :15.870
Mean   :15.140   Mean   :15.84   Mean   :16.21   Mean   :16.032
3rd Qu.:23.760   3rd Qu.:23.47   3rd Qu.:24.29   3rd Qu.:24.490
Max.   :34.630   Max.   :32.21   Max.   :32.54   Max.   :31.450
NA's   :4                                        NA's   :10
      MrdA                 Lern                Brgt                 Olit
Min.   : 0.125   Min.   : 1.042   Min.   : 0.125   Min.   : 1.562
1st Qu.: 8.180   1st Qu.: 8.080   1st Qu.: 8.180   1st Qu.: 8.680
Median :15.760   Median :15.210   Median :15.760   Median :16.770
Mean   :15.584   Mean   :15.531   Mean   :15.584   Mean   :16.627
3rd Qu.:23.065   3rd Qu.:23.480   3rd Qu.:23.065   3rd Qu.:24.590
Max.   :31.210   Max.   :32.110   Max.   :31.210   Max.   :32.400
NA's   :2                        NA's   :2
      Flcs                 Mrdf                Trbn                 Srtg
Min.   : 1.061   Min.   : 1.424   Min.   : 0.051   Min.   : 1.398
1st Qu.: 8.080   1st Qu.: 8.685   1st Qu.: 8.467   1st Qu.: 8.998
Median :15.810   Median :16.980   Median :18.030   Median :16.475
Mean   :15.952   Mean   :16.892   Mean   :17.609   Mean   :16.329
3rd Qu.:23.570   3rd Qu.:24.608   3rd Qu.:26.065   3rd Qu.:23.348
Max.   :32.220   Max.   :31.950   Max.   :33.250   Max.   :31.210
                 NA's   :13       NA's   :23       NA's   :27
```

```
        BR.P            Funs             BR.B             Cdrt
Min.    : 1.039   Min.    : 1.051   Min.    : 1.327   Min.    : 1.438
1st Qu.: 8.140   1st Qu.: 8.830   1st Qu.: 8.790   1st Qu.: 8.100
Median :16.660   Median :16.360   Median :16.990   Median :15.770
Mean   :16.540   Mean   :16.493   Mean   :16.465   Mean   :15.719
3rd Qu.:24.590   3rd Qu.:24.970   3rd Qu.:24.400   3rd Qu.:23.390
Max.   :32.010   Max.   :32.610   Max.   :32.270   Max.   :31.170


        Crll            Tudl             Fitr             Cscn
Min.    : 1.283   Min.    : 1.667   Min.    : 0.395   Min.    : 1.233
1st Qu.: 8.610   1st Qu.:10.043   1st Qu.: 7.850   1st Qu.: 8.140
Median :15.360   Median :17.375   Median :14.360   Median :15.080
Mean   :15.714   Mean   :17.453   Mean   :15.026   Mean   :15.751
3rd Qu.:23.100   3rd Qu.:25.317   3rd Qu.:22.950   3rd Qu.:23.640
Max.   :31.060   Max.   :30.880   Max.   :30.450   Max.   :32.130
                 NA's    :163
        Ablt            LsAr             Sesm
Min.    : 1.45   Min.    : 0.915   Min.    : 0.880
1st Qu.: 8.52   1st Qu.: 7.590   1st Qu.: 6.878
Median :15.51   Median :14.370   Median :13.230
Mean   :15.86   Mean   :15.044   Mean   :14.487
3rd Qu.:23.50   3rd Qu.:22.620   3rd Qu.:21.350
Max.   :31.05   Max.   :31.390   Max.   :31.280
                                  NA's    :24
```

6.2 Unemployment in the United States

As an example of a time series that can be displayed both in individual and in aggregate, we will use the unemployment data in the United States. The information on unemployed persons by industry and class of worker is available in Table A-14 published by the Bureau of Labor Statistics[3].

The dataset arranges the information with a row for each category (Series.ID) and a column for each monthly value. In addition, there are columns with the annual summaries (annualCols). We rearrange this data.frame, dropping the Series.ID and the annual columns, and transpose the data.

```
unemployUSA <- read.csv('data/TimeSeries/unemployUSA.csv')
nms <- unemployUSA$Series.ID
##columns of annual summaries
```

[3]http://www.bls.gov/webapps/legacy/cpsatab14.htm

95

```
annualCols <- 14 + 13*(0:12)
## Transpose. Remove annual summaries
unemployUSA <- as.data.frame(t(unemployUSA[,-c(1, annualCols)]))
## First 7 characters can be suppressed
names(unemployUSA) <- substring(nms, 7)
```

```
       32230                32231                32232                32235
Min.    :  2.00      Min.    : 384.0      Min.    : 596.0      Min.    : 701
1st Qu.: 22.00      1st Qu.: 626.8      1st Qu.: 774.2      1st Qu.:1019
Median : 30.00      Median : 823.0      Median :1024.0      Median :1176
Mean    : 37.76      Mean    : 981.0      Mean    :1091.4      Mean    :1303
3rd Qu.: 46.00      3rd Qu.:1190.2      3rd Qu.:1300.5      3rd Qu.:1707
Max.    :125.00      Max.    :2440.0      Max.    :2010.0      Max.    :2154
NA's    :6          NA's    :6          NA's    :6          NA's    :6
       32236                32237                32238                32239
Min.    :129.0      Min.    : 77.0      Min.    :184.0      Min.    : 504.0
1st Qu.:226.0      1st Qu.:144.0      1st Qu.:268.0      1st Qu.: 743.0
Median :267.0      Median :194.5      Median :317.0      Median : 923.5
Mean    :315.7      Mean    :200.9      Mean    :375.2      Mean    :1014.2
3rd Qu.:420.2      3rd Qu.:244.8      3rd Qu.:508.8      3rd Qu.:1346.8
Max.    :657.0      Max.    :373.0      Max.    :717.0      Max.    :1785.0
NA's    :6          NA's    :6          NA's    :6          NA's    :6
       32240                32241                32242                35109
Min.    : 293.0      Min.    : 636.0      Min.    :161.0      Min.    : 35.0
1st Qu.: 534.5      1st Qu.: 877.5      1st Qu.:265.0      1st Qu.:102.2
Median : 621.0      Median : 976.5      Median :313.5      Median :135.0
Mean    : 744.0      Mean    :1092.1      Mean    :351.6      Mean    :144.3
3rd Qu.: 994.8      3rd Qu.:1395.0      3rd Qu.:451.5      3rd Qu.:176.0
Max.    :1430.0      Max.    :1804.0      Max.    :618.0      Max.    :318.0
NA's    :6          NA's    :6          NA's    :6          NA's    :6
       28615                35181
Min.    : 269.0      Min.    :178.0
1st Qu.: 458.0      1st Qu.:260.5
Median : 543.5      Median :311.5
Mean    : 619.9      Mean    :371.8
3rd Qu.: 748.8      3rd Qu.:523.8
Max.    :1349.0      Max.    :730.0
NA's    :6          NA's    :6
```

With the transpose, the column names of the original data set are now the row names of the data.frame. The as.yearmon function of the zoo package converts the character vector of names into a yearmon vector, a class for representing monthly data. With Sys.setlocale("LC_TIME",

'C') we ensure that month abbreviations (%b) are correctly interpreted in a non-English locale. This vector is the time index of a new zoo object.

```
library("zoo")

Sys.setlocale("LC_TIME", 'C')
idx <- as.yearmon(row.names(unemployUSA), format = '%b.%Y')
unemployUSA <- zoo(unemployUSA, idx)
```

Finally, those rows with NA values are removed.

```
unemployUSA <- unemployUSA[complete.cases(unemployUSA), ]
```

```
    Index           32230            32231            32232
Min.   :2000   Min.   :  2.00   Min.   : 384.0   Min.   : 596.0
1st Qu.:2003   1st Qu.: 22.00   1st Qu.: 626.8   1st Qu.: 774.2
Median :2006   Median : 30.00   Median : 823.0   Median :1024.0
Mean   :2006   Mean   : 37.76   Mean   : 981.0   Mean   :1091.4
3rd Qu.:2009   3rd Qu.: 46.00   3rd Qu.:1190.2   3rd Qu.:1300.5
Max.   :2012   Max.   :125.00   Max.   :2440.0   Max.   :2010.0
    32235           32236            32237            32238
Min.   : 701   Min.   :129.0    Min.   : 77.0    Min.   :184.0
1st Qu.:1019   1st Qu.:226.0    1st Qu.:144.0    1st Qu.:268.0
Median :1176   Median :267.0    Median :194.5    Median :317.0
Mean   :1303   Mean   :315.7    Mean   :200.9    Mean   :375.2
3rd Qu.:1707   3rd Qu.:420.2    3rd Qu.:244.8    3rd Qu.:508.8
Max.   :2154   Max.   :657.0    Max.   :373.0    Max.   :717.0
    32239           32240            32241            32242
Min.   : 504.0   Min.   : 293.0   Min.   : 636.0   Min.   :161.0
1st Qu.: 743.0   1st Qu.: 534.5   1st Qu.: 877.5   1st Qu.:265.0
Median : 923.5   Median : 621.0   Median : 976.5   Median :313.5
Mean   :1014.2   Mean   : 744.0   Mean   :1092.1   Mean   :351.6
3rd Qu.:1346.8   3rd Qu.: 994.8   3rd Qu.:1395.0   3rd Qu.:451.5
Max.   :1785.0   Max.   :1430.0   Max.   :1804.0   Max.   :618.0
    35109           28615            35181
Min.   : 35.0    Min.   : 269.0   Min.   :178.0
1st Qu.:102.2    1st Qu.: 458.0   1st Qu.:260.5
Median :135.0    Median : 543.5   Median :311.5
Mean   :144.3    Mean   : 619.9   Mean   :371.8
3rd Qu.:176.0    3rd Qu.: 748.8   3rd Qu.:523.8
Max.   :318.0    Max.   :1349.0   Max.   :730.0
```

6.3   Gross National Income and CO_2 Emissions

The catalog data of the World Bank Open Data initiative includes the
World Development Indicators (WDI)[4]. Among them, we will analyze the
Gross National Income (GNI) for a set of countries from 2000 to 2021.

```r
library("WDI")

countries <- c('CN', 'DE', 'ES',
               'FR', 'GR', 'IN',
               'NO', 'US')

## GNI, PPP (current international $)
## GNI per capita, PPP (current international $)
## Population
WDIdata <- WDI(indicator = c("NY.GNP.MKTP.PP.CD",
                             "NY.GNP.PCAP.PP.CD",
                             "SP.POP.TOTL"),
               start = 2000, end = 2021,
               country = countries)

names(WDIdata) <- c("Country.Name", "iso2c", "Country.Code",
                    "Year",
                    "GNI.PPP", "GNI.capita", "Population")
```

Country.Name	iso2c	Country.Code	Year
Length:176	Length:176	Length:176	Min. :2000
Class :character	Class :character	Class :character	1st Qu.:2005
Mode :character	Mode :character	Mode :character	Median :2010
			Mean :2010
			3rd Qu.:2016
			Max. :2021

GNI.PPP	GNI.capita	Population
Min. :1.638e+11	Min. : 2070	Min. :4.491e+06
1st Qu.:7.724e+11	1st Qu.:20108	1st Qu.:3.321e+07
Median :2.655e+12	Median :31870	Median :7.406e+07
Mean :5.451e+12	Mean :33149	Mean :3.885e+08
3rd Qu.:7.252e+12	3rd Qu.:47085	3rd Qu.:5.135e+08
Max. :2.862e+13	Max. :89010	Max. :1.414e+09

---

[4]http://databank.worldbank.org/data/reports.aspx?source=
world-development-indicators

On the other hand, the data explorer of the Climate Watch website publishes the historical values of CO_2 emissions[5], and can be downloaded as a CSV file.

```
## Bulk data downloaded from CW
## MtCO2e of CO2, Total including LUCF
CWdata <- read.csv2("data/TimeSeries/CO2_ClimateWatch.csv")
```

These two data.frames are to be merged together. First, we filter CWdata to preserve the same countries that are contained in WDIdata.

```
## Filter by countries
countries3 <- unique(WDIdata$Country.Code)
CWdata <- subset(CWdata,
                 Country %in% countries3)
```

Then, it is transformed from wide to long format with the function melt of the package reshape2.

```
library("reshape2")

CWdata <- melt(CWdata,
               id.vars = "Country",
               variable.name = "Year",
               value.name = "CO2")

## Remove the X in year values
CWdata$Year <- as.numeric(substring(CWdata$Year, 2, 5))
```

Now they can be merged together, and a new variable is calculated, CO2.capita.

```
CO2data <- merge(WDIdata, CWdata,
                 by.x = c("Country.Code", "Year"),
                 by.y = c("Country", "Year"))

CO2data$CO2.capita <- with(CO2data,
                           CO2 / Population * 1e6)
```

Finally, the Country.Code and Country.Name variables are converted to factors, which will be useful with some graphics.

```
CO2data$Country.Code <- factor(CO2data$Country.Code)
CO2data$Country.Name <- factor(CO2data$Country.Name)
```

[5]https://www.climatewatchdata.org/data-explorer/historical-emissions

99

```
 Country.Code        Year         Country.Name      iso2c
CHN    :22      Min.    :2000    China  :22    Length:176
DEU    :22      1st Qu.:2005    France :22    Class :character
ESP    :22      Median :2010    Germany:22    Mode  :character
FRA    :22      Mean    :2010    Greece :22
GRC    :22      3rd Qu.:2016    India  :22
IND    :22      Max.    :2021    Norway :22
(Other):44                       (Other):44
      GNI.PPP            GNI.capita        Population            CO2
Min.   :1.638e+11  Min.   : 2070    Min.   :4.491e+06  Min.   :    15.51
1st Qu.:7.724e+11  1st Qu.:20108    1st Qu.:3.321e+07  1st Qu.:   167.38
Median :2.655e+12  Median :31870    Median :7.406e+07  Median :   471.87
Mean   :5.451e+12  Mean   :33149    Mean   :3.885e+08  Mean   :  1928.00
3rd Qu.:7.252e+12  3rd Qu.:47085    3rd Qu.:5.135e+08  3rd Qu.:  2566.51
Max.   :2.862e+13  Max.   :89010    Max.   :1.414e+09  Max.   :10849.18

   CO2.capita          CO2.PPP           cexPoints
Min.   : 0.6762   Min.   :0.04652   Min.   :0.5000
1st Qu.: 4.1302   1st Qu.:0.13574   1st Qu.:0.5000
Median : 5.6616   Median :0.24272   Median :0.9333
Mean   : 6.7633   Mean   :0.27305   Mean   :0.8964
3rd Qu.: 8.9374   3rd Qu.:0.34425   3rd Qu.:0.9333
Max.   :19.0211   Max.   :0.83932   Max.   :1.8000
```

Part II

Spatial Data

Chapter 7

Displaying Spatial Data: Introduction

Spatial data (also known as geospatial data) are directly or indirectly referenced to a location on the surface of the Earth. Their spatial reference is composed of coordinate values and a system of reference for these coordinates. Spatial data are often accessed, manipulated, or analyzed through Geographic Information Systems (GIS).

Real objects represented by GIS data can be divided into two abstractions: discrete objects (e.g., a road or a river) represented with vector data (points, lines, and polygons), and continuous fields (such as elevation or solar radiation) represented with raster data. The sp and sf packages are the preferred option to use vector data in R, and the raster and terra packages are the choice for raster data [1].

This part exposes several examples where vector and raster data are displayed. These examples make use of several datasets (available at the book website) described in Chapter 15.

On the one hand, Chapters 8, 9, 10, and 11 focus on thematic maps that display a specific variable, commonly using geographic data such as coastlines, boundaries, and places as points of reference for the variable being

---

[1] Although sp, sf, raster, and terra are the most important packages, there are an increasing number of packages designed to work with spatial data. They are summarized in the corresponding CRAN Task View. Read Section 7.2 for details.

DOI: 10.1201/9781003485384-7

mapped. These maps provide specific information about particular locations or areas (proportional symbol mapping, choropleth, and cartogram maps) and information about spatial patterns (isarithmic and raster maps).

On the other hand, Chapter 12 is devoted to vector fields, Chapter 14 focuses on reference maps to show the geographic location of features, and Chapter 13 deals with physical maps to show the landscape and features of a place.

7.1 Packages

The CRAN Tasks View "Analysis of Spatial Data" [2] summarizes the packages for reading, visualizing, and analyzing spatial data. This section provides a brief introduction to sp, sf, raster, terra, rasterVis, and gstat. Most of the information has been extracted from their vignettes, webpages, and help pages. You should read them for detailed information.

7.1.1 sp

The sp package (E. J. Pebesma and Bivand 2005) provides classes and methods for dealing with spatial data in R. The spatial data classes implemented are points (SpatialPoints), grids (SpatialPixels and SpatialGrid), lines (Line, Lines, and SpatialLines), rings, and polygons (Polygon, Polygons, and SpatialPolygons), each of them without data or with data (for example, SpatialPointsDataFrame or SpatialLinesData-Frame)[3].

Selecting, retrieving, or replacing certain attributes in spatial objects with data is done using standard methods:

- [selects rows (items) and columns in the data.frame.

- [[selects a column from the data.frame

- [[<- assigns or replaces values to a column in the data.frame.

A number of spatial methods are available for the classes in sp:

---

[2] http://CRAN.R-project.org/view=Spatial

[3] The asterisk is commonly used as a wildcard character to denote subsets of classes. Thus, SpatialLines* comprises SpatialLines and SpatialLinesDataFrame classes. Moreover, Spatial* represents all the classes defined by the sp package.

- `coordinates(object) <- value` sets spatial coordinates to create spatial data. It promotes a `data.frame` into a `SpatialPointsData-Frame`. *value* may be specified by a formula, a character vector, or a numeric matrix or `data.frame` with the actual coordinates.

- `coordinates(object, ...)` returns a matrix with the spatial coordinates. If used with `SpatialPolygons` it returns a matrix with the centroids of the polygons.

- `bbox` returns a matrix with the coordinates of the bounding box.

- `proj4string(object)` and `proj4string(object) <- value` retrieve or set projection attributes on spatial classes.

- `spTransform` transforms from one coordinate reference system (geographic projection) to another.

- `spplot` plots attributes combined with spatial data: points, lines, grids, and polygons.

7.1.2   sf

The `sf` package (E. Pebesma 2024), the long term successor of `sp`, implements simple features in R. Simple features is an open (OGC and ISO) interface standard for access and manipulation of spatial vector data (points, lines, polygons).

This package represents simple features using simple data structures, commonly `data.frame` objects. Feature geometries are stored in a data.frame column, using a list-column because geometries are not single-valued. The length of this list is equal to the number of records in the `data.frame`, with the simple feature geometry of that feature in each element of the list.

`sf` implements three classes to represent simple features:

- `sf`, a `data.frame` with feature attributes and feature geometries. It contains

- `sfc`, the list-column with the geometries for each feature (record), which is composed of

- `sfg`, the feature geometry of an individual simple feature.

All functions and methods in sf that operate on spatial data are pre-
fixed by st_ (spatial and temporal). For the purposes of this book, the most
important are:

- st_read, st_write, for reading and writing spatial data, respectively.

- st_transform for coordinate reference system transformations.

- st_as_sf.*, a family of conversion functions between sp and sf.

The sf package implements plot methods for displaying data using
base graphics. Besides, this package provides a number of methods for
conversion to grob objects in order to display these objects with packages
working with the grid system (lattice and ggplot2). Finally, the ggplot2
package contains the geom_sf geom, designed for sf objects.

7.1.3   raster

The raster package (R. J. Hijmans 2024a) has functions for creating, reading, manipulating, and writing raster data. The package provides general
raster data manipulation functions. The package also implements raster
algebra and most functions for raster data manipulation that are common
in Geographic Information Systems (GIS).

The raster package can work with raster datasets stored on disk if
they are too large to be loaded into memory. The package can work with
large files because the objects it creates from these files only contain information about the structure of the data, such as the number of rows and
columns, the spatial extent, and the filename, but it does not attempt to
read all the cell values in memory. In computations with these objects, the
data are processed in chunks.

The package defines a number of S4 classes. RasterLayer, Raster-
Brick, and RasterStack are the most important:

- A RasterLayer object represents single-layer (variable) raster data. It
  can be created with the function raster. This function is able to create a RasterLayer from another object, including another Raster*
  object[4], or a SpatialPixels* and SpatialGrid* object, or even a
  matrix. In addition, it can create a RasterLayer reading data from

---

[4]The notation Raster* represents all the classes of Raster objects: RasterLayer,
RasterStack, and RasterBrick.

a file. The `raster` package can use raster files in several formats. Supported formats for reading include GeoTIFF, ESRI, ENVI, and ERDAS.

- `RasterBrick` and `RasterStack` are classes for multilayer data. A `RasterStack` is a list of `RasterLayer` objects with the same spatial extent and resolution. A `RasterStack` can be formed with a collection of files in different locations or even mixed with `RasterLayer` objects that only exist in memory. A `RasterBrick` is truly a multilayered object, and processing it can be more efficient than processing a `RasterStack` representing the same data.

The `raster` package defines a number of methods for raster algebra with `Raster*` objects: arithmetic operators, logical operators, and functions such as `abs`, `round`, `ceiling`, `floor`, `trunc`, `sqrt`, `log`, `log10`, `exp`, `cos`, `sin`, `max`, `min`, `range`, `prod`, `sum`, `any`, and `all`. In these functions, `Raster*` objects can be mixed with numbers.

There are several functions to modify the content or the spatial extent of `Raster*` objects or to combine `Raster*` objects:

- The `crop` function takes a geographic subset of a larger `Raster*` object. `trim` crops a `RasterLayer` by removing the outer rows and columns that only contain NA values. `extend` adds new rows and/or columns with NA values.

- The `merge` function merges two or more `Raster*` objects into a single new object.

- `projectRaster` transforms values of a `Raster*` object to a new object with a different coordinate reference system.

- With `overlay`, multiple `Raster*` objects can be combined (for example, multiply them).

- `mask` removes all values from one layer that are NA in another layer, and `cover` combines two layers by taking the values of the first layer except where these are NA.

- `calc` computes a function for a `Raster*` object. With `RasterLayer` objects, another `RasterLayer` is returned. With multilayer objects, the result depends on the function: With a summary function (`sum`, `max`, etc.), `calc` returns a `RasterLayer` object, and a `RasterBrick` object otherwise.

- stackApply computes summary layers for subsets of a RasterStack or RasterBrick.

- cut and reclassify replace ranges of values with single values.

- zonal computes zonal statistics, that is, summarizes a Raster* object using zones (areas with the same integer number) defined by another RasterLayer.

7.1.4 terra

The terra package (R. J. Hijmans 2024b) is the successor of the raster package. Both packages share a similar interface, although, according to the author, "terra is simpler, faster, and can do more".

terra implements two main classes: SpatRaster for raster data and SpatVector for spatial vector data. This is a significant difference from the raster package, which defines three different classes for raster data and six classes for spatial data.

The main functions and methods in terra are the same as in raster, with minor changes in their names[5]. These are the most relevant:

- The crop function takes a geographic subset of a larger SpatRaster object. trim crops a SpatRaster by removing the outer rows and columns that only contain NA values. extend adds new rows and/or columns with NA values.

- The union function merges two or more SpatRaster objects into a single new object.

- project transforms values of a SpatRaster object to a new object with a different coordinate reference system.

- mask and cover removes values or combines layers depending on NA cells.

- app computes a function for a SpatRaster object, tapp computes summary layers, and lapp applies a function to a SpatRaster object using layers as arguments.

- classify replace ranges of values with single values.

- zonal computes zonal statistics.

---

[5]The package webpage includes a section with a comparison between the raster and terra functions and methods.

7.1.5 rasterVis

The rasterVis package (O. Perpiñán and R. Hijmans 2023) complements the raster and terra packages, providing a set of methods for enhanced visualization and interaction. This package defines visualization methods (levelplot) for quantitative data and categorical data, both for univariate and multivariate rasters.

It also includes several methods in the frame of the Exploratory Data Analysis approach: scatterplots with xyplot, histograms and density plots with histogram and densityplot, violin and boxplots with bwplot, and a matrix of scatterplots with splom.

Finally, this package is able to display vector fields using arrows, vectorplot, or with streamlines (Wegenkittl and Gröller 1997), streamplot.

7.1.6 gstat

The gstat package (E. J. Pebesma 2004) provides functions for geostatistical modeling, prediction, and simulation, including variogram modeling and simple, ordinary, universal, and external drift kriging.

Most of the functionality of this package is beyond the scope of this book. However, some functions must be mentioned:

- variogram calculates the sample variogram from data or for the residuals if a linear model is given. vgm generates a variogram and fit.variogram fit ranges and/or sills from a variogram model to a sample variogram.

- krige is the function for simple, ordinary, or universal kriging. gstat is the function for univariate or multivariate geostatistical prediction.

7.2 Further Reading

- (Slocum 2022) and (Dent, Torguson, and Hodler 2008) are comprehensive books on thematic cartography and geovisualization. They include chapters devoted to data classification, scales, map projections, color theory, typography, and proportional symbol, choropleth and cartogram, dasymetric, isarithmic, and multivariate mapping. Several resources are available at the accompanying website [6].

---

[6]https://www.routledge.com/Thematic-Cartography-and-Geovisualization/
Slocum-McMaster-Kessler-Howard/p/book/9780367712709

- (Bivand, E. J. Pebesma, and Gomez-Rubio 2013) is the essential reference to work with spatial data in R. R. Bivand and E. Pebesma are the authors of the fundamental sp package, and they are the authors or maintainers of several important packages such as gstat, for geostatistical modeling, prediction, and simulation. Chapter 3 is devoted to the visualization of spatial data. Code, figures, and data of the book are available at the accompanying website [7].

- (Hengl 2009) is an open-access book with seven spatial data analysis exercises.

- The CRAN Tasks View "Analysis of Spatial Data" [8] summarizes the packages for reading, visualizing, and analyzing spatial data. The packages in development published at R-Forge are listed in the "Spatial Data & Statistics" topic view [9]. The R-SIG-Geo mailing list [10] is a powerful resource for obtaining help.

---

[7] http://www.asdar-book.org/
[8] http://CRAN.R-project.org/view=Spatial
[9] http://r-forge.r-project.org/softwaremap/trove_list.php?form_cat=353
[10] https://stat.ethz.ch/mailman/listinfo/R-SIG-Geo/

Chapter 8

Thematic Maps: Proportional Symbol Mapping

The proportional symbol technique uses symbols of different sizes to represent data associated with areas or point locations, with circles being the most frequently used geometric symbol. The data and the size of symbols can be related through different types of scaling: mathematical scaling sizes areas of point symbols in direct proportion to the data; perceptual scaling corrects the mathematical scaling to account for visual underesstimation of larger symbols; and range grading, where data are grouped, and each class is represented with a single symbol size.

The most relevant packages used in this chapter are: sp and sf for reading and writing spatial data; classInt for computing class intervals; gstat for spatial interpolation; and mapview and rgl for interactive visualization.

DOI: 10.1201/9781003485384-8

8.1 Introduction

In this chapter we display data from a grid of sensors belonging to the Integrated Air Quality system of the Madrid City Council (Section 15.1) with circles as the proportional symbol and range grading as the scaling method. The objective when using range grading is to discriminate between classes instead of estimating an exact value from a perceived symbol size. However, because human perception of symbol size is limited, it is always recommended to add a second perception channel to improve the discrimination task. Colors from a sequential palette will complement symbol size to encode the groups.

Two alternatives are available to display the data:

1. The `sp` package and the `spplot` function (based on `lattice` graphics).

2. The `ggplot2` package and the `geom_sf` function.

```
library("sp")
library("sf")
```

In both cases, the data is imported with the `sf` package and its function `st_read`. This function returns an object of class `sf`.

```
NO2sf <- st_read(dsn = "data/Spatial/", layer = "NO2sf")
```

The data will be represented with a sequential palette defined with `colorRampPalette`. Colors denote the value of the variable: green for lower values of the contaminant, brown for intermediate values, and black for the highest.

```
airPal <- colorRampPalette(c("springgreen1", "sienna3", "gray5"))
    (5)
```

8.2 Proportional Symbol Mapping

The `ggplot2` package is able to display spatial point observations with the `geom_sf` function. This function understands the classes defined by the `sf` package that provides the function `st_read` to read the data:

```
## Create a categorical variable
NO2sf$Mean <- cut(NO2sf$mean, 5)
```

Both color and size can be combined in a unique graphical output using aes and `scale_fill_manual`:

```
ggplot(data = NO2sf) +
    geom_sf(aes(size = Mean, fill = Mean),
            pch = 21, col = 'black') +
    scale_fill_manual(values = airPal) +
    theme_bw()
```

The `spplot` method provided by the `sp` package, based on `xyplot` from the `lattice` package, is able to display `SpatialPointsDataFrame` objects, among other classes. The `sf` object returned by `st_read` has to be converted to a `Spatial` class.

```
NO2sp <- as(NO2sf, "Spatial")
```

The arguments `col.regions` and `cex` in `spplot` are used for encoding color and size (Figure 8.1).

```
spplot(NO2sp["mean"],
       col.regions = airPal, ## Palette
       cex = sqrt(1:5), ## Size of circles
       edge.col = "black", ## Color of border
       scales = list(draw = TRUE), ## Draw scales
       key.space = "right") ## Put legend on the right
```

8.3 Optimal Classification to Improve Discrimination

Two main improvements can be added to Figure 8.1:

- Define classes dependent on the data structure (instead of the uniform distribution assumed with cut). A suitable approach is the `classInterval` function of the `classInt` package, which implements the Fisher-Jenks optimal classification algorithm[1]. This classification method seeks to reduce the variance within classes and maximize the variance between classes.

```
library("classInt")
## The number of classes is chosen between the Sturges and the
## Scott rules.
nClasses <- 5
```

---

[1]This classification method will be used in section 9.2 with a choropleth map.

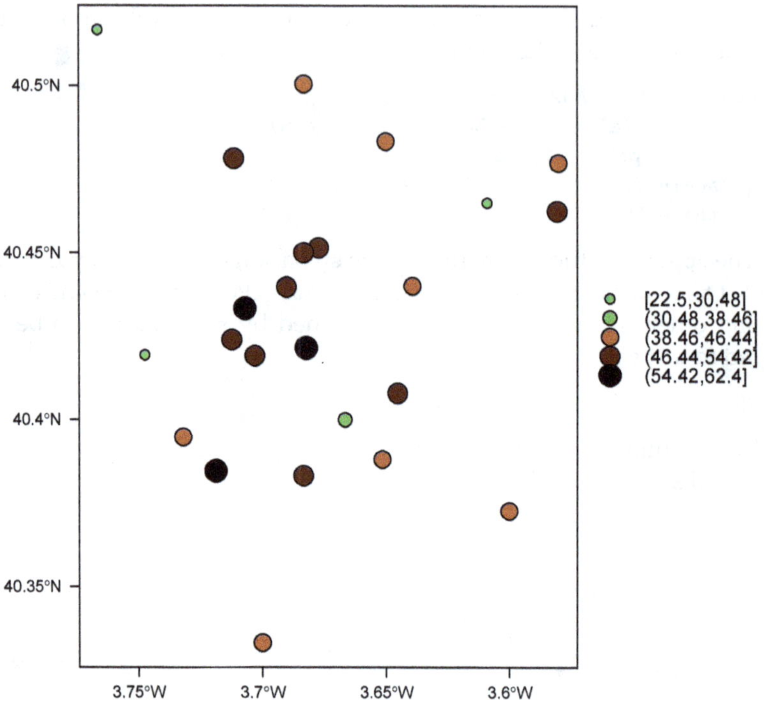

FIGURE 8.1: Annual average of NO_2 measurements in Madrid. Values are shown with different symbol sizes and colors for each class with the sp-plot function.

```
intervals <- classIntervals(NO2sp$mean, n = nClasses, style = "
    fisher")
## Number of classes is not always the same as the proposed
    number
nClasses <- length(intervals$brks) - 1

op <- options(digits = 4)
tab <- print(intervals)
options(op)
```

- Encode each group with a symbol size (circle area) such that visual discrimination among classes is enhanced. The next code uses the set

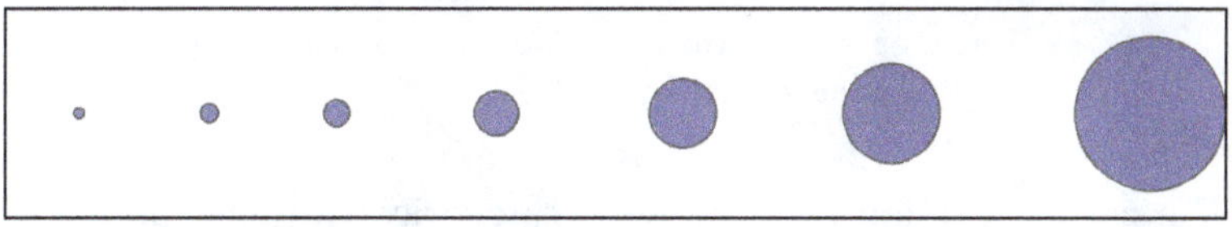

FIGURE 8.2: Symbol sizes proposed by Borden Dent.

of radii proposed in (Dent, Torguson, and Hodler 2008) (Figure 8.2). This set of circle sizes is derived from studies by Meihoefer (Meihoefer 1969). He derived a set of ten circle sizes that were easily and consistently discriminated by his subjects. The alternative proposed by Dent et al. improves the discrimination between some of the circles.

```
## Complete Dent set of circle radii (mm)
dent <- c(0.64, 1.14, 1.65, 2.79, 4.32, 6.22, 9.65, 12.95, 15.11)
## Subset for our dataset
dentAQ <- dent[seq_len(nClasses)]
## Link Size and Class: findCols returns the class number of each
## point; cex is the vector of sizes for each data point
idx <- findCols(intervals)
cexNO2 <- dentAQ[idx]
```

These two enhancements are included in Figure 8.3, which displays the categorical variable `classNO2` (instead of mean), whose levels are the intervals previously computed with `classIntervals`. In addition, this figure includes an improved legend.

```
## spplot version
NO2sp$classNO2 <- factor(names(tab)[idx])

## Definition of an improved key with title and background
NO2key <- list(x = 0.99, y = 0.01, corner = c(1, 0),
               title = expression(NO[2]~~(paste(mu, plain(g))/m^3)),
               cex.title = 0.8, cex = 1,
               background = "gray92")

pNO2 <- spplot(NO2sp["classNO2"],
               col.regions = airPal,
               cex = dentAQ * 0.8,
               edge.col = "black",
```

```
               scales = list(draw = TRUE),
               key.space = NO2key)
     pNO2
```

The ggplot version uses the same categorical variable, added to the NO2sf object.

```
## ggplot2 version
NO2sf$classNO2 <- factor(names(tab)[idx])

ggplot(data = NO2sf) +
    geom_sf(aes(size = classNO2, fill = classNO2),
            pch = 21, col = "black") +
    scale_fill_manual(values = airPal) +
    scale_size_manual(values = dentAQ * 2) +
    xlab("") + ylab("") + theme_bw()
```

8.4 Spatial Context with Underlying Layers and Labels

The spatial distribution of the stations is better understood if we add underlying layers with information about the spatial context.

A common solution is to add an underlying layer with a static image representing a map. Such an image can be downloaded from a provider such as Google Maps™ or OpenStreetMap. There are packages such as ggmap that provide an interface to these map servers, providing raster images from static maps.

This approach faces two main problems: the user can neither modify the image nor use its content to produce additional information; most of the providers require registration and payments and impose usage restrictions due to copyright.

A different approach is to use digital vector data (points, lines, and polygons) that can be filtered and transformed. I will illustrate this approach with two different solutions: remote data obtained from the OpenStreetMap project with the osmdata package; local data contained in a shapefile (a format commonly used by public and private providers to distribute information).

8.4.1 OpenStreetMap with osmdata

The osmdata package is able to access the OpenStreepMap data with queries defined with the functions opq and add_osm_feature. The opq

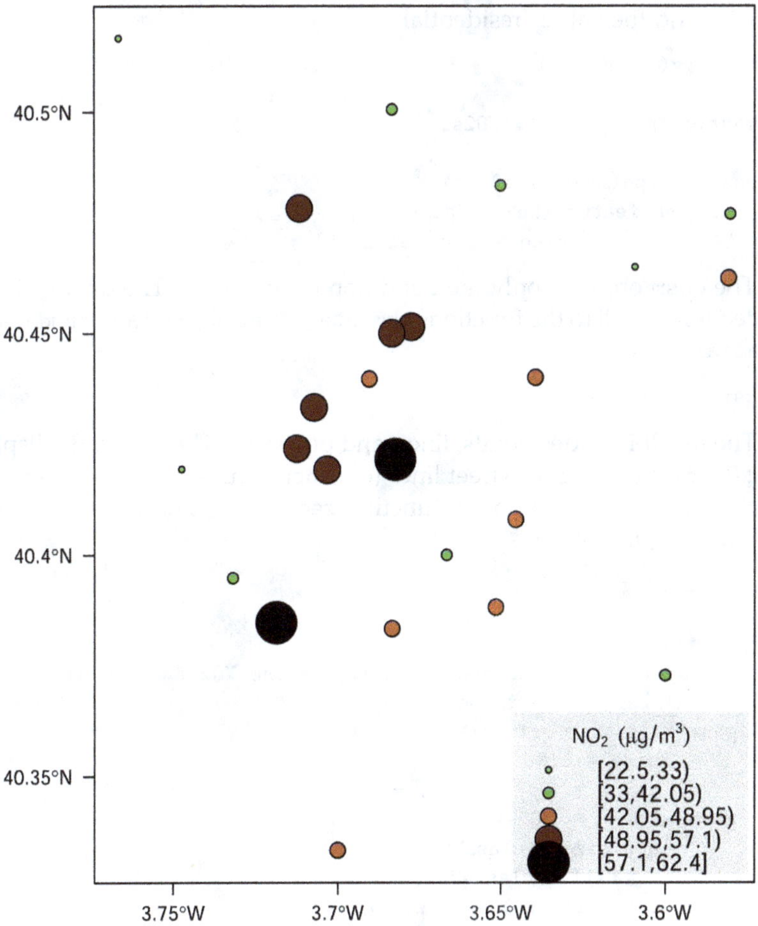

FIGURE 8.3: Annual average of NO_2 measurements in Madrid. Enhancement of Figure 8.1, using symbol sizes proposed by Borden Dent and an improved legend.

function defines the base query (region of interest in this example), and the `add_osm_feature` filters the features to be downloaded. The streets in a city can be retrieved with the key "highway" (any kind of road, street, or path) and the value "residential"[2].

```
library("osmdata")

madridBox <- st_bbox(NO2sf)

qosm <- opq(madridBox) %>%
  add_osm_feature(key = "highway",
                  value = "residential")
```

The qosm object is only the definition of the query. The data is downloaded with a call to the functions `osmdata_sf` (sf objects) and `osmdata_sp` (Spatial* objects).

```
qsf <- osmdata_sf(qosm)
```

The result includes points, lines, and polygons. The next code displays the NO2sf data using the street lines (qsf$osm_lines) as the context and a label for each station with the function geom_text_repel of the ggrepel package (Figure 8.4).

```
library("ggrepel")

ggplot()+
  ## Layers are drawn sequentially, so the NO2sf layer must be in
  ## the last place to be on top
  geom_sf(data = qsf$osm_lines["name"],
          size = 0.3,
          color = "lightgray") +
  geom_sf(data = NO2sf,
          aes(size = classNO2,
              fill = classNO2),
          pch = 21, col = "black") +
  ## Labels for each point, with position according to the circle
      size
  ## and the rest of labels
  geom_text_repel(data = NO2sf,
                  aes(label = substring(codEst, 7),
                      geometry = geometry,
                      point.size = classNO2),
```

---

[2]More details are available at https://wiki.openstreetmap.org/wiki/Key:highway.

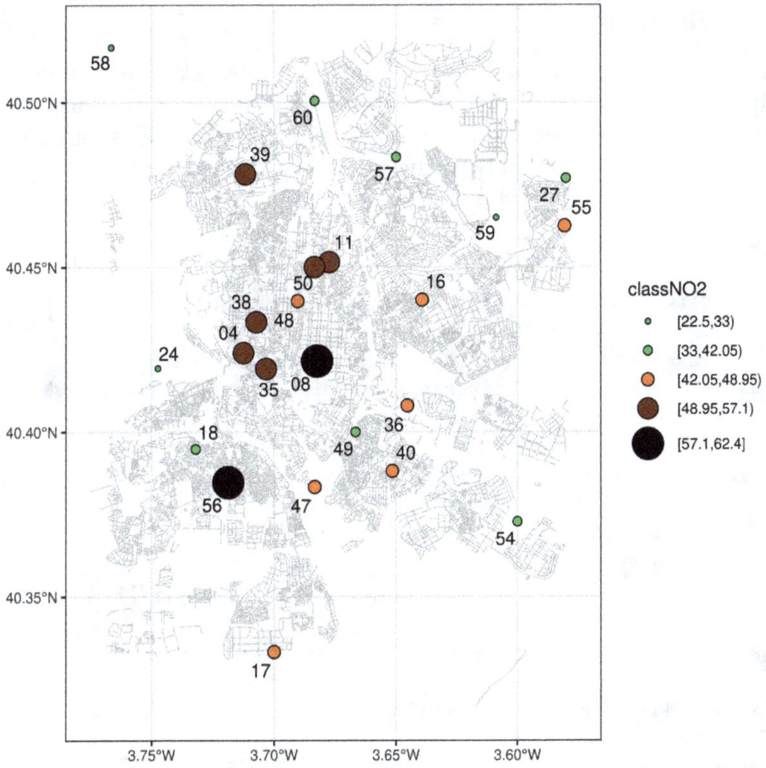

FIGURE 8.4: Annual average of NO_2 measurements in Madrid. Enhancement of Figure 8.3, displaying OpenStreepMap data in an underlying layer.

```
                    stat = "sf_coordinates") +
    scale_fill_manual(values = airPal) +
    scale_size_manual(values = dentAQ * 2) +
    labs(x = NULL, y = NULL) + theme_bw()
```

The `spplot` version needs the data to be downloaded as `Spatial*` objects with `osmdata_sp`.

```
qsp <- osmdata_sp(qosm)
```

The result can be combined with the `NO2sp` object with the `sp.layout` argument of `spplot`, which accepts a call to a function defined in a list.

119

8.4.2 Shapefiles

The geographical information of Madrid is available at the "nomecalles" web service[3]. The data folder contains compressed files with several shapefiles downloaded from this service. These shapefiles can be read with the st_read function from the sf package.

```
## Madrid districts
unzip("data/Spatial/distr2022.zip", exdir = tempdir())

distritosMadridSF <- st_read(dsn = tempdir(),
                             layer = "dist2022")
## Filter the streets of the Municipality of Madrid
distritosMadridSF <- distritosMadridSF[distritosMadridSF$CMUN ==
    "079",]
## Assign the geographical reference
distritosMadridSF <- st_transform(distritosMadridSF,
                             crs = "WGS84")

## Madrid streets
unzip("data/Spatial/call2022.zip", exdir = tempdir())

streetsMadridSF <- st_read(dsn = tempdir(),
                           layer = "Gdie_g_calles")
streetsMadridSF <- streetsMadridSF[streetsMadridSF$CDMUNI == "079
    ",]
streetsMadridSF <- st_transform(streetsMadridSF,
                           crs = "WGS84")
```

The sf objects produced with these code can be displayed together with the NO2sf object with consecutive calls to geom_sf. Figure 8.5 displays the final result.

```
ggplot()+
  geom_sf(data = streetsMadridSF,
          size = 0.1,
          color = "darkgray") +
  geom_sf(data = distritosMadridSF,
          fill = "lightgray",
          alpha = 0.2,
          size = 0.15,
          color = "black") +
  geom_sf(data = NO2sf,
```

[3]https://gestiona.comunidad.madrid/nomecalles_web/

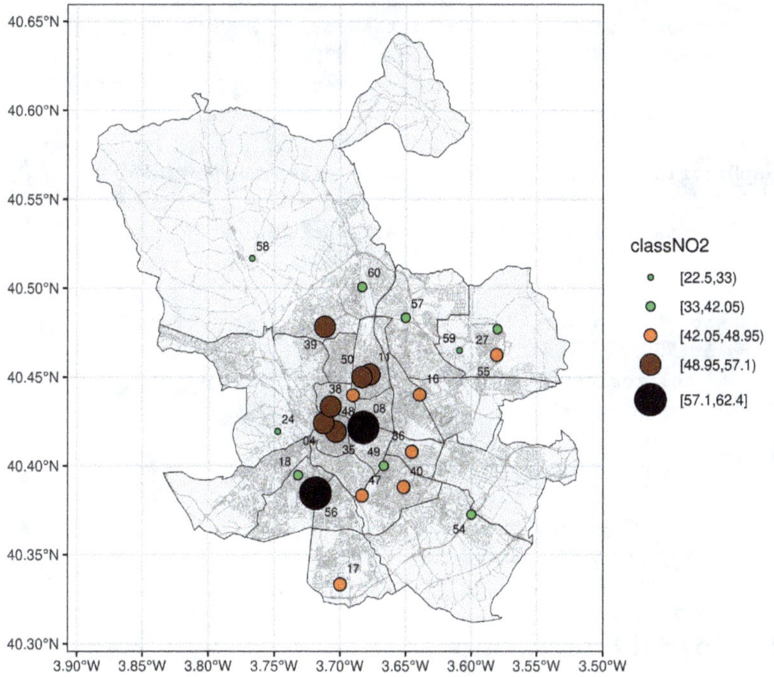

FIGURE 8.5: Annual average of NO_2 measurements in Madrid using shapefiles (lines and polygons) and text as geographical context.

```
            aes(size = classNO2,
                fill = classNO2),
            pch = 21, col = "black") +
    geom_text_repel(data = NO2sf,
                    aes(label = substring(codEst, 7),
                        geometry = geometry,
                        point.size = classNO2),
                    size = 2.5,
                    stat = "sf_coordinates") +
    scale_fill_manual(values = airPal) +
    scale_size_manual(values = dentAQ * 2) +
    labs(x = NULL, y = NULL) + theme_bw()
```

These shapefiles can also be used with spplot (after a class conversion) thanks to the sp.layout argument of this function or with the layer and +.trellis functions from the latticeExtra package.

```
distritosMadridSP <- as(distritosMadridSF, "Spatial")
streetsMadridSP <- as(streetsMadridSF, "Spatial")
## Lists using the structure accepted by sp.layout, with the
    polygons,
## lines, and points, and their graphical parameters
spDistricts <- list("sp.polygons", distritosMadridSP,
                    fill = "gray97", lwd = 0.3)
spStreets <- list("sp.lines", streetsMadridSP,
                  lwd = 0.05)

## spplot with sp.layout version
spplot(NO2sp["classNO2"],
       col.regions = airPal,
       cex = dentAQ,
       edge.col = "black",
       alpha = 0.8,
       sp.layout = list(spDistricts, spStreets),
       scales = list(draw = TRUE),
       key.space = NO2key)
```

8.5 Spatial Interpolation

The measurements at discrete points give limited information about the underlying process. It is quite common to approximate the spatial distribution of the measured variable with the interpolation between measurement locations. Selection of the optimal interpolation method is outside the scope of this book. The interested reader is referred to (Cressie and C. Wikle 2015) and (Bivand, E. J. Pebesma, and Gomez-Rubio 2013).

The following code illustrates an easy solution using inverse distance weighted (IDW) interpolation with the gstat package (E. J. Pebesma 2004) *only* for illustration purposes.

```
library("gstat")

## Sample 10^5 points locations within the bounding box of NO2sp
    using
## regular sampling
airGrid <- spsample(NO2sp, type = "regular", n = 1e5)
## Convert the SpatialPoints object into a SpatialGrid object
gridded(airGrid) <- TRUE
## Compute the IDW interpolation
airKrige <- krige(mean ~ 1, NO2sp, airGrid)
```

The result is a `SpatialPixelsDataFrame` that can be displayed with
`spplot` and combined with the previous layers and the measurement station points (Figure 8.6).

```
spplot(airKrige["var1.pred"], ## Variable interpolated
      col.regions = colorRampPalette(airPal)) +
  layer({ ## Overlay boundaries and points
    sp.polygons(distritosMadridSP,
                fill = "transparent",
                lwd = 0.3)
    sp.lines(streetsMadridSP,
             lwd = 0.07)
    sp.points(NO2sp,
              pch = 21,
              alpha = 0.8,
              fill = "gray50",
              col = "black")
})
```

8.6 Interactive Graphics

Now, let's suppose you need to know the median and standard deviation
of the time series of a certain station. Moreover, you would like to view the
photography of that station, or even better, you wish to visit its webpage
for additional information. A frequent solution is to produce interactive
graphics with tooltips and hyperlinks.

In this section we visit several approaches to create these products: the
`mapview` package based on the `htmlwidgets` package; export to GeoJSON
and KML formats; and 3D visualization with the `rgl` package.

8.6.1 mapview

The syntax of `mapview`[4] resembles the syntax of `spplot`. Its first argument
is the spatial object with the information, and the variable to be depicted
is selected with the argument `zcol`. Moreover, the size of the points can
be linked to another variable with the argument `cex`, and their labels extracted from another variable with the argument `label`.

The next code produces an HTML page with an interactive graphic
inserted in it (Figure 8.7). When the mouse is hovered over a point, its

---

[4]The package `mapview` is able to work both with `sp` and `sf` objects. In this section the
code works with `sp` objects but would work without modification with `sf` objects.

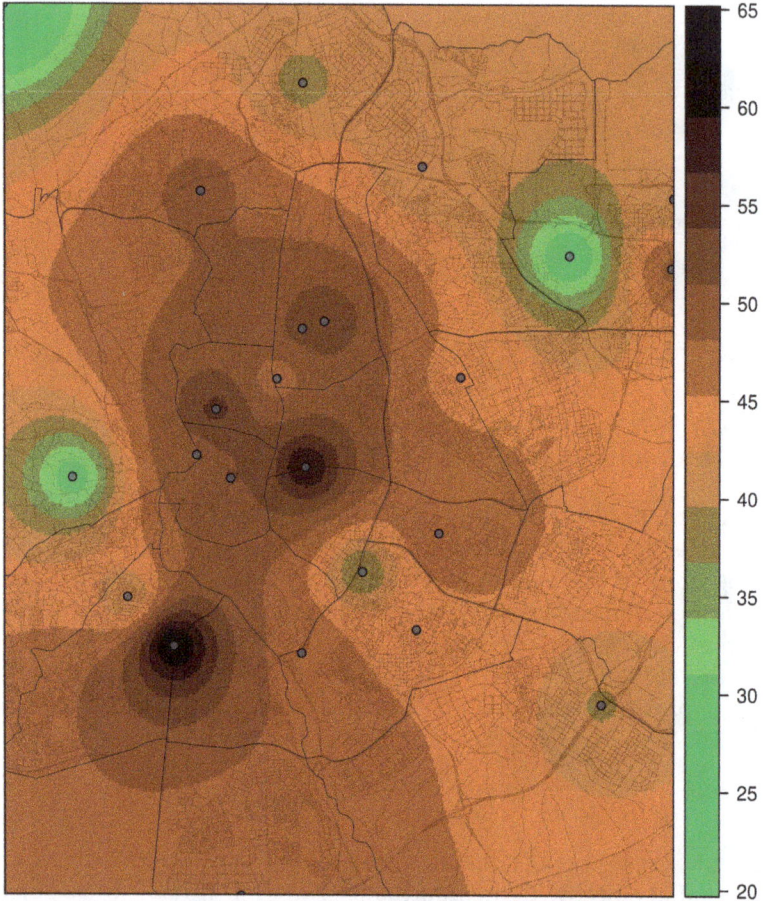

FIGURE 8.6: Kriging annual average of NO_2 measurements in Madrid.

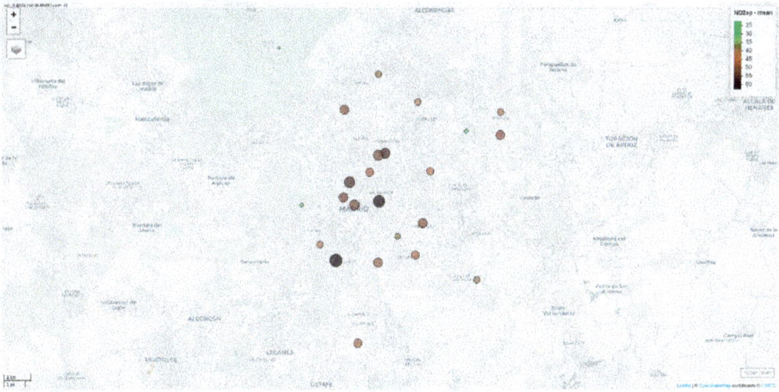

FIGURE 8.7: Snapshot of the interactive graphic produced with `mapview` depicting the annual average of NO_2 measurements in Madrid.

label is displayed, and if the point is selected, a tooltip with the whole information is deployed.

```
library("mapview")

pal <- colorRampPalette(c("springgreen1", "sienna3", "gray5"))

mapview(NO2sp,
        zcol = "mean", ## Variable to display
        cex = "mean", ## Use this variable for the circle sizes
        col.regions = pal,
        label = NO2sp$Nombre,
        legend = TRUE)
```

8.6.1.1   Tooltips with images and graphs

The tooltip included in the previous graphic is very simple: only text displaying a table with information. This tooltip can be improved thanks to the popup argument and the popup* family of functions. For example, the next code creates an interactive graphic whose tooltips show an image of the station (available in the `images` folder of the repository[5]) using the `popupImage` function (Figure 8.8).

---

[5]These images have been obtained from the Air Quality portal of Madrid, https://airedemadrid.madrid.es/, following the path "Actuaciones municipales > Sistema integral de calidad del aire > Vigilancia".

FIGURE 8.8: Snapshot of the interactive graphic produced with `mapview` with tooltips including images.

As an additional feature, the provider[6] of the background map is selected with the argument `map.type`.

```r
library("leafpop")

img <- paste("images/Spatial/", NO2sp$codEst, ".jpg", sep = "")

mapview(NO2sp,
    zcol = "mean",
    cex = "mean",
    col.regions = pal,
    label = NO2sp$Nombre,
    popup = popupImage(img, src = "local", embed = TRUE),
    map.type = "Esri.WorldImagery",
    legend = TRUE)
```

A more sophisticated solution displays a scatterplot when a tooltip is deployed. The `popupGraph` function accepts a list of graphics and selects the one corresponding to the location selected by the user. This list is produced with the next code: first, the measurements time series is read and filtered; second, the stations code is extracted; finally, a loop with `lapply` creates a time series graphic for each station displaying the evolution of the measurements along the time period.

---

[6]The list of providers is available in http://leaflet-extras.github.io/ leaflet-providers/preview/

126

```
## Read the time series
airQuality <- read.csv2("data/Spatial/airQuality.csv")
## We need only NO2 data (codParam 8)
NO2 <- subset(airQuality, codParam == 8)
## Time index in a new column
NO2$tt <- with(NO2,
               as.Date(paste(year, month, day, sep = "-")))
## Stations code
stations <- unique(NO2$codEst)
## Loop to create a scatterplot for each station.
pList <- lapply(stations,
               function(i)
                   xyplot(dat ~ tt, data = NO2,
                       subset = (codEst == i),
                       type = "l",
                       xlab = "", ylab = "")
               )
```

This list of graphics, pList, is provided to mapview through the popup argument with the function popupGraph (Figure 8.9).

```
mapview(NO2sp,
        zcol = "mean",
        cex = "mean",
        col.regions = pal,
        label = NO2sp$Nombre,
        popup = popupGraph(pList),
        map.type = "Esri.WorldImagery",
        legend = TRUE)
```

8.6.1.2 Synchronise multiple graphics

The mapview package recreates the small multiple technique (Sections 3.2 and 4.1) with the functions sync and latticeView. With them, multiple variables can be rendered simultaneously and synchronized together (with the sync function):

- if a panel is zoomed, all other panels will also zoom

- the mouse position in a panel is signaled with a red circle in the rest of the panels.

The next code generates three graphics to view different variables of the NO2sp object using different values in zcol and cex. All of them are viewed and synchronized together with sync (Figure 8.10):

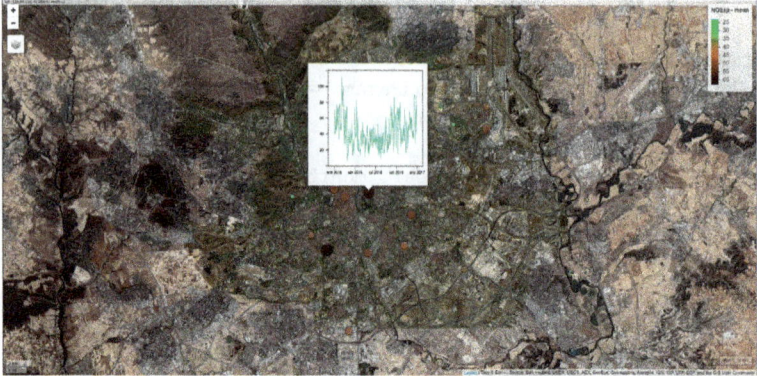

FIGURE 8.9: Snapshot of the interactive graphic produced with `mapview` with tooltips including time series graphics.

```r
library("leafsync")

## Map of the average value
mapMean <- mapview(NO2sp, zcol = "mean", cex = "mean",
                   col.regions = pal, legend = TRUE,
                   map.types = "OpenStreetMap.Mapnik",
                   label = NO2sp$Nombre)

## Map of the median
mapMedian <- mapview(NO2sp, zcol = "median", cex = "median",
                     col.regions = pal, legend = TRUE,
                     #map.type = "NASAGIBS.ViirsEarthAtNight",
                     label = NO2sp$Nombre)

## Map of the standard deviation
mapSD <- mapview(NO2sp, zcol = "sd", cex = "sd",
                 col.regions = pal, legend = TRUE,
                 map.type = "Esri.WorldImagery",
                 label = NO2sp$Nombre)

## All together
sync(mapMean, mapMedian, mapSD, ncol = 3)
```

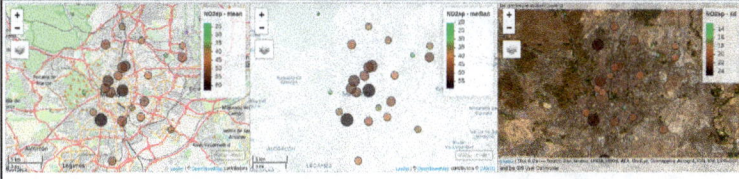

FIGURE 8.10: Snapshot of multiple interactive graphics produced with `mapview`.

8.6.2 Export to Other Formats

A different approach is to use an external data viewer due to its features or its large community of users. Two tools deserve to be mentioned: GeoJSON rendered within GitHub repositories and Keyhole Markup Language (KML) files imported in Google Earth™.

8.6.2.1 GeoJSON and OpenStreetMap

GeoJSON is an open computer file format for encoding collections of simple geographical features along with their nonspatial attributes using JavaScript Object Notation (JSON). The `sf` package can produce GeoJSON files with the `st_write` function.

```
st_write(NO2sf,
        dsn = "data/Spatial/NO2.geojson",
        layer = "NO2sp",
        driver = "GeoJSON")
```

Figure 8.11 shows a snapshot of the rendering of this GeoJSON file using the geojson.io[7] service. You can zoom on the map and click on the stations to display the data.

8.6.2.2 Keyhole Markup Language

Keyhole Markup Language (KML) is a file format to display geographic data within Internet-based, two-dimensional maps and three-dimensional Earth browsers. KML uses a tag-based structure with nested elements and attributes and is based on the XML standard. KML became an international standard of the Open Geospatial Consortium in 2008. Google Earth

---

[7]https://geojson.io/

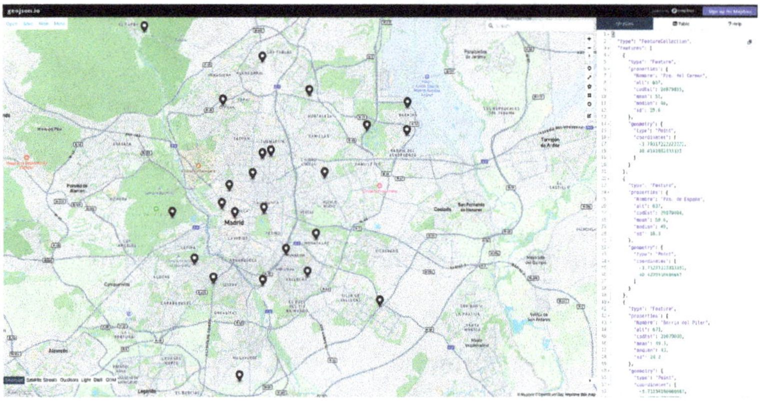

FIGURE 8.11: NO_2 data in a GeoJSON file rendered with the geojson.io service.

was the first program able to view and graphically edit KML files, although Marble, an open-source project, also offers KML support.

There are several packages able to generate KML files. For example, the `st_write` function from the `sf` package can also write KML files:

```
st_write(NO2sf,
        dsn = "data/Spatial/NO2_mean.kml",
        layer = "mean",
        driver = "KML")
```

This file can be directly opened with Google Earth or Marble.

8.6.3 3D visualization

An alternative method is 3D visualization, where the user can rotate or zoom the figure. This solution is available thanks to the `rgl` package, which provides functions for 3D interactive graphics.

```
library("rgl")
```

Previously, the `SpatialPointsDataFrame` object must be converted to a `data.frame`. The `xyz` coordinates will be the longitude, latitude, and altitude of each station.

```
## rgl does not understand Spatial* objects
NO2df <- as.data.frame(NO2sp)
```

The color of each point is determined by the corresponding class (Section 8.3), and the radius of each bubble depends on the mean value of the depicted variable.

```
## Color of each point according to its class
airPal <- colorRampPalette(c("springgreen1", "sienna3", "gray5"))
    (5)
colorClasses <- airPal[NO2df$classNO2]
```

A snapshot of this graphic is displayed in Figure 8.12.

```
plot3d(x = NO2df$coords.x1,
    y = NO2df$coords.x2,
    z = NO2df$alt,
    xlab = "Longitude",
    ylab = "Latitude",
    zlab = "Altitude",
    type = "s",
    col = colorClasses,
    radius = NO2df$mean/10)
```

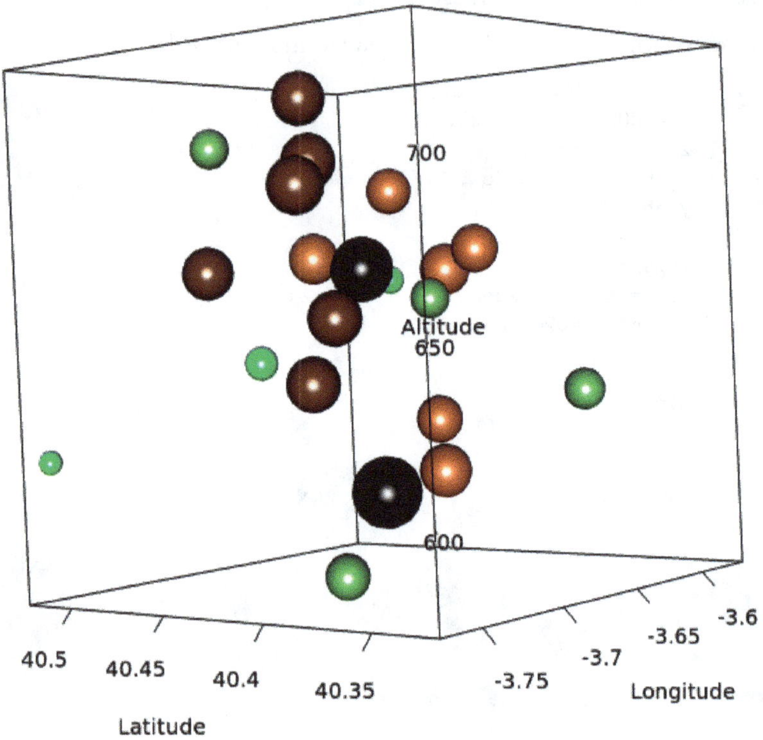

FIGURE 8.12: Snapshot of the interactive graphic produced with rgl.

Chapter 9

Thematic Maps: Choropleth Maps

A choropleth map shades regions according to the measurement of a variable displayed on the map. The choropleth map is an appropriate tool to visualize a variable uniformly distributed within each region, changing only at the region boundaries. This method performs correctly with homogeneous regions, both in size and shape. The next subjects are covered in this chapter: sequential and qualitative palettes; small multiples; class intervals; and interactive visualization.

The most relevant packages used in this chapter are: sp and sf for reading and writing spatial data; classInt for computing class intervals; RColorBrewer for color palettes; and mapview for interactive visualization.

9.1 Introduction

This chapter details how to create choropleth maps depicting the results of the 2016 Spanish general elections. Section 15.2 describes how to define a shapefile[1] by combining the data from a `data.frame` and the spatial information of the administrative boundaries.

As exposed in Chapter 8, two alternatives are available:

1. The `sp` package and the `spplot` function (based on `lattice` graphics).

2. The `sf` package and the `geom_sf` function (based on `ggplot2` graphics).

```
library("sp")

library("sf")
```

In both cases, the data is imported with the `sf` package and its function `st_read`, returning an object of class `sf`.

9.1.1 Read Data

The shapefile is read with the `st_read` to produce an `sf` object. Because the coordinate reference system[2] is not stored in the files, it must be set with the `crs` argument.

```
sfMapVotes <- st_read("data/Spatial/sfMapVotes.shp")
```

This `object` contains two main variables: `whichMax`, the name of the predominant political option, and `pcMax`, the percentage of votes obtained by this political option.

```
     SP_ID               PROVMUN             whichMax
 Length:8110         Length:8110         Length:8110
 Class :character    Class :character    Class :character
 Mode  :character    Mode  :character    Mode  :character
```

---

[1]The shapefile format is a geospatial vector data format developed by Esri that can store vector features (points, lines, and polygons) and attributes that describe these features (name of the location, temperature, etc.).

[2]The EPSG projection of the data is 25830. More information at https://spatialreference.org/ref/epsg/25830/.

```
        Max                 pcMax                  geometry
 Min.   :      2.0    Min.    :21.33   MULTIPOLYGON :8110
 1st Qu.:     54.0    1st Qu.:31.68    epsg:25830   :   0
 Median :    162.0    Median :35.64    +proj=utm ...:   0
 Mean   :   1395.9    Mean    :37.58
 3rd Qu.:    636.5    3rd Qu.:41.27
 Max.   :696804.0     Max.    :94.74
```

9.1.2 Province Boundaries

As a visual aid, the subsequent maps will be produced with the province boundaries superimposed. These boundaries are encoded in the `spain_-provinces` shapefile.

```
sfProvs <- st_read("data/Spatial/spain_provinces.shp",
                   crs = 25830)
```

9.2 Quantitative Variable

First, let's display `pcMax`, a quantitative variable increasing from low to high. This type of variable is well suited to sequential palettes that communicate the progression from low to high with light colors associated with low values and dark colors linked to high values. The well-known service ColorBrewer[3] provides several choices, available in R via the `RColorBrewer` package (Neuwirth 2022).

```
## Number of intervals (colors)
N <- 6
## Sequential palette
quantPal <- brewer.pal(n = N, "Oranges")
```

The most common approach with choropleth maps, the classed choropleth, is to divide the data into classes. Although this method produces a filtered view of the data, it reduces the random noise in the information, and makes it easy to compare regions. A different alternative is the unclassed choropleth, where each unique data value gets a unique color. This approach is recommended to get an unfiltered view of the data and highlight overall geographic patterns of the variable.

Figure 9.1 is an unclassed choropleth depicting the `pcMax` variable. It uses a palette created by interpolation with the `colorRampPalette` function.

---

[3]http://colorbrewer2.org

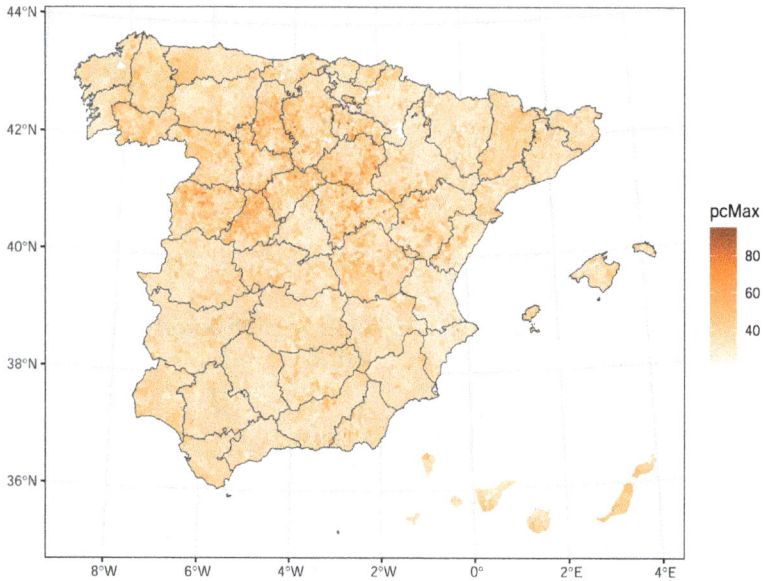

FIGURE 9.1: Quantitative choropleth map displaying the percentage of votes obtained by the predominant political option in each municipality in the 2016 Spanish general elections using a continuous color ramp (unclassed choropleth).

```
ggplot(sfMapVotes) +
  ## Display the pcMax variable...
  geom_sf(aes(fill = pcMax),
          ## without drawing municipality boundaries
          color = "transparent") +
  scale_fill_gradientn(colours = quantPal) +
  ## And overlay provinces boundaries
  geom_sf(data = sfProvs,
          fill = 'transparent',
          ## but do not include them in the legend
          show.legend = FALSE) +
  theme_bw()
```

The `spplot` version requires a conversion from `sf` to a `Spatial-PolygonsDataFrame` class with `as`.

```
spMapVotes <- as(sfMapVotes, "Spatial")
spProvs <- as(sfProvs, "Spatial")
```

The `spMapVotes` object is displayed with `spplot` and the province boundaries are superimposed with the argument `sp.layout` (see section 8.4.1 for details).

```
## Number of cuts
ucN <- 1000
## Palette created with interpolation
ucQuantPal <- colorRampPalette(quantPal)(ucN)

## Province boundaries
provinceLines <- list("sp.polygons",
                      spProvs,
                      lwd = 0.1,
                      # draw the lines after the data
                      first = FALSE)

## Main plot
spplot(spMapVotes["pcMax"],
       col.regions = ucQuantPal,
       cuts = ucN,
       ## Do not draw municipality boundaries
       col = "transparent",
       ## Overlay province boundaries
       sp.layout = provinceLines)
```

9.2.1 Data Classification

It is evident in Figure 9.1 that the `pcMax` variable is concentrated in the 0.2-0.4 range. Figure 9.2 displays the density estimation of this variable grouping by the political option. This result suggests to use data classification.

```
ggplot(as.data.frame(spMapVotes),
       aes(pcMax,
           fill = whichMax,
           colour = whichMax)) +
    geom_density(alpha = 0.1) +
    theme_bw()
```

The number of data classes is the result of a compromise between information amount and map legibility. A general recommendation is to use three to seven classes, depending on the data.

On the other hand, there is a wide catalog of classification methods, and the `classInt` package implements most of them (previously used in

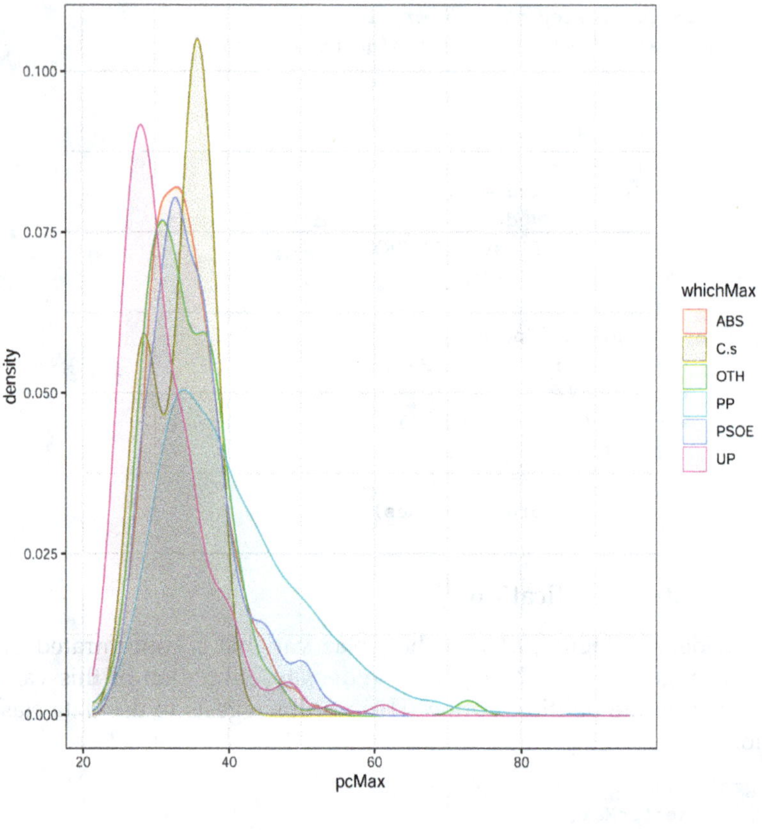

FIGURE 9.2: Density estimation of the predominant political option in each municipality in the 2016 Spanish general elections, grouped by the political option.

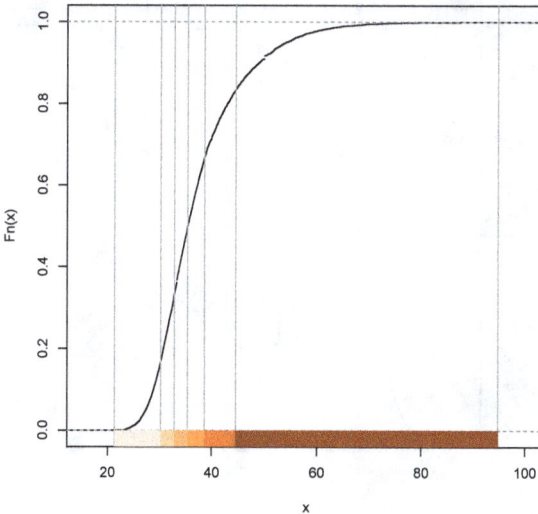

FIGURE 9.3: Quantile method for setting class intervals.

Section 8.3). Figures 9.3 and 9.4 depict the empirical cumulative distribution function of pcMax with the intervals computed with the quantile method and the natural breaks method, a clustering method that seeks to reduce the variance within classes and maximize the variance between classes. As it can be inferred from the density estimation (Figure 9.2), the natural breaks method is preferred in this example, because with the quantile method very different values will be assigned to the same class.

```r
library("classInt")

## Compute intervals with the same number of elements
intQuant <- classIntervals(sfMapVotes$pcMax,
                    n = N, style = "quantile")
## Compute intervals with the natural breaks algorithm
intFisher <- classIntervals(sfMapVotes$pcMax,
                    n = N, style = "fisher")

plot(intQuant, pal = quantPal, main = "")

plot(intFisher, pal = quantPal, main = "")
```

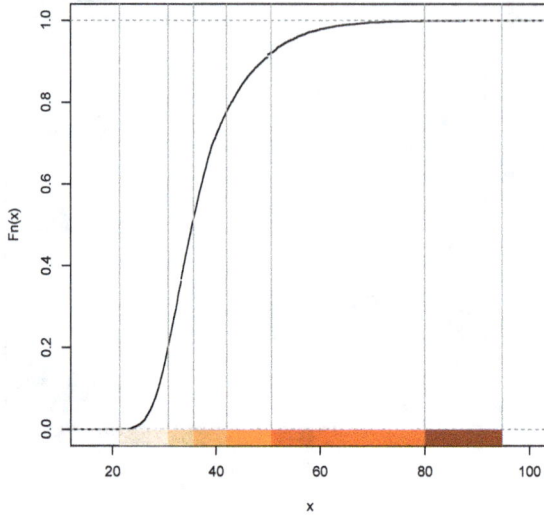

FIGURE 9.4: Natural breaks method for setting class intervals.

Figure 9.5 is a classed choropleth with the natural breaks classification. It is produced with spplot displaying a categorical variable created with the function cut and the breaks computed with classIntervals.

```
## spplot solution

## Add a new categorical variable with cut, using the computed
   breaks
spMapVotes$pcMaxInt <- cut(spMapVotes$pcMax,
                         breaks = intFisher$brks,
                         include.lowest = TRUE)

spplot(spMapVotes["pcMaxInt"],
     col = "transparent",
     col.regions = quantPal,
     sp.layout = provinceLines)
```

The next code is the version of this figure with ggplot2:

```
## sf and geom_sf
sfMapVotes$pcMaxInt <- cut(sfMapVotes$pcMax,
                         breaks = intFisher$brks,
                         include.lowest = TRUE)
```

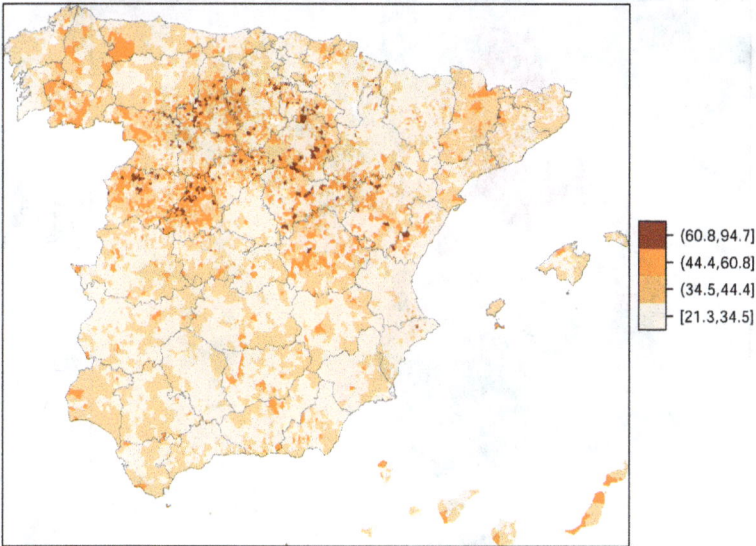

FIGURE 9.5: Quantitative choropleth map displaying the percentage of votes obtained by the predominant political option in each municipality in the 2016 Spanish general elections using a classification (classed choropleth).

```
ggplot(sfMapVotes) +
  geom_sf(aes(fill = pcMaxInt),
          color = "transparent") +
  scale_fill_brewer(palette = "Oranges") +
  geom_sf(data = sfProvs,
          fill = "transparent",
          show.legend = FALSE) +
  theme_bw()
```

9.3 Qualitative Variable

On the other hand, whichMax is a categorical value with four levels: the main parties (PP, PSOE, UP, Cs), the abstention results (ABS), and the rest of the parties (OTH).

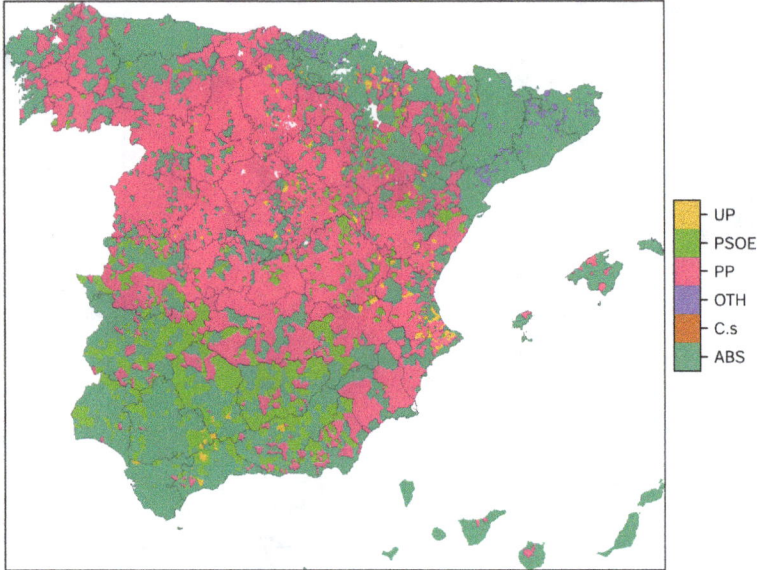

FIGURE 9.6: Categorical choropleth map displaying the name of the predominant political option in each municipality in the 2016 Spanish general elections.

```
 ABS  C.s  OTH   PP PSOE   UP
2812    3  170 4212  776  137
```

Figure 9.6 encodes these levels with a qualitative palette from Color-Brewer.

```
classes <- levels(spMapVotes$whichMax)
nClasses <- length(classes)

qualPal <- brewer.pal(nClasses, "Dark2")

## spplot solution
spplot(spMapVotes["whichMax"],
       col.regions = qualPal,
       col = 'transparent',
       sp.layout = provinceLines)
```

The next code is the version of this figure with ggplot2:

```
## geom_sf solution
ggplot(sfMapVotes) +
  geom_sf(aes(fill = whichMax),
          color = "transparent") +
  scale_fill_brewer(palette = "Dark2") +
  geom_sf(data = sfProvs,
          fill = "transparent",
          show.legend = FALSE) +
  theme_bw()
```

9.4   Small Multiples with Choropleth Maps

Both the quantitative and qualitative variables can be combined using the small multiples technique (Sections 3.2 and 4.1) (Tufte 1990): multiple maps displayed all at once to compare the differences between them. The next code produces a matrix of maps, with a map for each political option defined by the categorical variable whichMax. The spplot function provides a formula argument to divide the data into panels. However, its usage is not well documented and cannot be recommended. Instead, the ggplot approach is easy to use thanks to the facet_wrap function. The result is displayed in Figure 9.7.

```
ggplot(sfMapVotes) +
  geom_sf(aes(fill = pcMaxInt),
          color = "transparent") +
  ## Define the faceting using two rows
  facet_wrap(~whichMax, nrow = 2) +
  scale_fill_brewer(palette = "Oranges") +
  geom_sf(data = sfProvs,
          fill = "transparent",
          size = 0.1,
          show.legend = FALSE) +
  theme_bw()
```

9.5   Bivariate Map

Following the inspiring example of the infographic titled "Immigration Explorer" published by the *New York Times*[4], we will combine the

---

[4]Available at https://archive.nytimes.com/www.nytimes.com/interactive/2009/03/10/us/20090310-immigration-explorer.html.

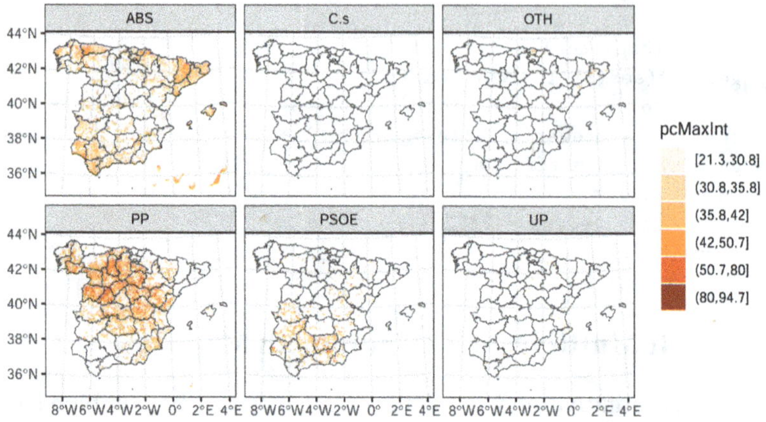

FIGURE 9.7: Small multiple choropleth maps of the Spanish general election results. Each map shows the results of a political option in each municipality.

choropleth maps of both variables to produce a bivariate map[5]: the hue of each polygon will be determined by the name of the predominant option (whichMax), but the transparency will vary according to the percentage of votes (pcMax).

In previous sections, we use six intervals to represent the quantitative variable pcMax. However, in this case we must reduce this number: in order to improve the map legibility, each ramp has only four steps.

Moreover, the qualitative variable whichMax will also reduce, will also go from six to 4, grouping the political parties into political options:

```
## PP and Cs -> Right
## PSOE and UP -> Left
levels(sfMapVotes$whichMax) <-
    c("ABS", "Right", "OTH", "Right", "Left", "Left")
```

The n code creates a bidimensional palette joining four different sequential palettes (one for each level of whichMax) with four steps. Thus, the bivariate legend will be composed of sixteen colors.

---

[5]Although bivariate maps are generally used to display the relationship between two variables, they can also be used to display one variable and its uncertainty. More information about visualizing uncertainty with maps can be found in (Lucchesi and C. K. Wikle 2017) and the package VizU (https://github.com/pkuhnert/VizU).

```
## Number of steps.
Nint <- 4
## ABS - Greys, Right - Blues, OTH - Greens, Left - Reds
multiPal <- lapply(c("Greys", "Blues", "Greens", "Reds"),
                function(pal) brewer.pal(Nint, pal))
multiPal <- do.call(rbind, multiPal)
```

The `biscale` package contains functions for bivariate mapping with ggplot2. First, the function `bi_class` returns a `sf` object with a new column, `bi_class`, defining the class of each element according to the value of `whicMax` and `pcMax`.

```
library("biscale")


sfClass <- bi_class(sfMapVotes,
            x = whichMax,
            y = pcMax,
            style = "fisher",
            dim = 4)
```

The next step is to create the legend with `bi_legend`. This function requires a bivariate palette as a named vector. The next code transforms the previous `multiPal` matrix into a vector, adds the required names, and produces the legend.

```
bipal <- c(multiPal)

nms <- outer(1:4, 1:4, paste, sep = "-")
names(bipal) <- c(nms)

bilegend <- bi_legend(pal = bipal,
                    dim = 4,
                    xlab = "ABS-Right-OTH-Left",
                    ylab = "% of votes ",
                    size = 8)
```

Finally, the bivariate map is created with `geom_sf` combined with `bi_-scale_fill`, both of them fed with the last results, `sfClass` and `bipal`. However, the legend cannot be added directly.

```
bimap <- ggplot() +
  geom_sf(data = sfClass,
        aes(fill = bi_class),
        color = "white",
```

```
        size = 0.1,
        show.legend = FALSE) +
  bi_scale_fill(pal = bipal, dim = 4) +
  bi_theme()
```

The complete map is produced with the package cowplot and its functions ggdraw and draw_plot. This function displays the map (bimap) and the legend (bilegend) in their respective locations. Figure 9.8 displays the result.

```
library("cowplot")

ggdraw() +
  draw_plot(bimap, 0, 0, 1, 1) +
  draw_plot(bilegend, 0.05, 0.1,
            width = 0.2, height = 0.2)
```

There is no package for the spplot approach, so the solution must be built step by step. First, the classes are defined with classIntervals and cut:

```
## Define the intervals
intFisher <- classIntervals(spMapVotes$pcMax,
                    n = Nint, style = "fisher")
## ... and create a categorical variable with them
spMapVotes$pcMaxInt <- cut(spMapVotes$pcMax,
                    breaks = intFisher$brks)
```

Then, we can produce a list of maps, extracting the polygons according to each class of the qualitative variable and filling them with the appropriate color from the multiPal palette. The resulting list of trellis objects can be combined with Reduce and the +.trellis function of the latticeExtra and produce a trellis object.

```
levels(spMapVotes$whichMax) <-
  c("ABS", "Right", "OTH", "Right", "Left", "Left")

classes <- levels(spMapVotes$whichMax)
nClasses <- length(classes)

pList <- lapply(1:nClasses, function(i)
{
  ## Only those polygons corresponding to a level are selected
  mapClass <- subset(spMapVotes,
                    whichMax == classes[i])
```

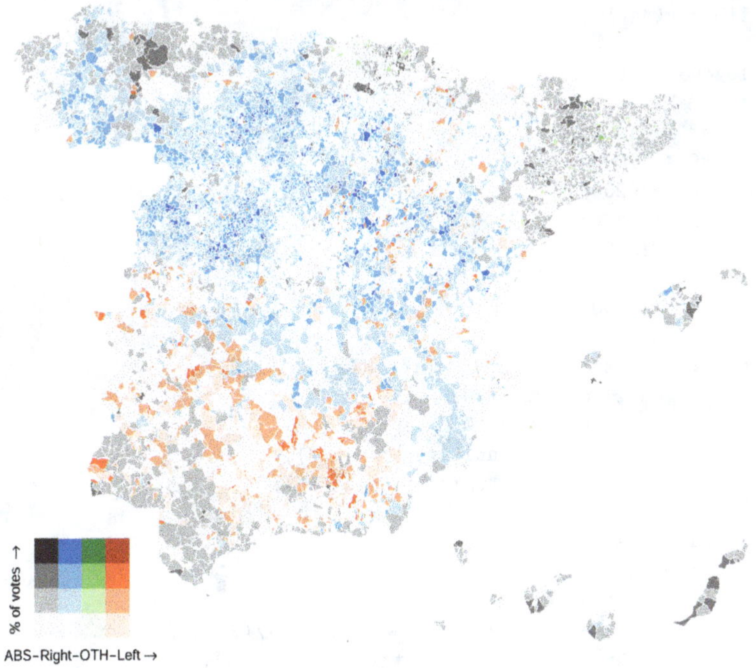

FIGURE 9.8: Bidimensional choropleth map of the Spanish general election results. The map shows the result of the most voted option in each municipality.

```
## Palette
pal <- multiPal[i, ]
## Produce the graphic
pClass <- spplot(mapClass, "pcMaxInt",
                 col.regions = pal,
                 col = "transparent",
                 colorkey = FALSE)
})
names(pList) <- classes
p <- Reduce("+", pList)
```

The bidimensional legend of this graphic is produced with `grid.raster`, a function of the `grid` package, able to display a color matrix (line 13). The axis of the color matrix are created with `grid.text` (lines 19 and 30).

```r
library("grid")

legend <- layer(
{
  ## Position of the legend
  x0 <- 1000000
  y0 <- 4200000
  ## Width of the legend
  w <- 120000
  ## Height of the legend
  h <- 100000
  ## Colors
  grid.raster(multiPal, interpolate = FALSE,
              x = unit(x0, "native"),
              y = unit(y0, "native"),
              width = unit(w, "native"),
              height = unit(h, "native"))
  ## x-axis (quantitative variable)
  Ni <- length(intervals)
  grid.text(intervals,
            y = unit(y0 - 1.25 * h/2, "native"),
            x = unit(seq(x0 - w * (Ni -1)/(2*Ni),
                     x0 + w * (Ni -1)/(2*Ni),
                     length = Ni),
                  "native"),
            just = "top",
            rot = 45,
            gp = gpar(fontsize = 6))
  ## y-axis (qualitative variable)
  grid.text(classes,
            y = unit(seq(y0 + h * (nClasses -1)/(2*nClasses),
                     y0 - h * (nClasses -1)/(2*nClasses),
                     length = nClasses),
                  "native"),
            x = unit(x0 + w/2, "native"),
            just = "left",
            gp = gpar(fontsize = 6))
})
```

Last step, the bivariate map and the legend are displayed together:

```r
## Main plot
p + legend
```

9.6 Interactive Graphics

The package `mapview` was used in section 8.6.1 to produce interactive proportional symbol maps. In this section, this package creates interactive choropleth maps.

```
library("mapview")
```

This package is able to work both with `sp` and with `sf`. In this section we use the `sf` package to read the data[6].

```
sfMapVotes0 <- st_read("data/Spatial/sfMapVotes0.shp",
                crs = 25830)
```

Figures 9.9 and 9.10 show the snapshots of the interactive choropleth maps of `pcMax` and `whichMax`, respectively. These maps are produced with the next code.

```
## Quantitative variable, pcMax
mapView(sfMapVotes0,
        zcol = "pcMax", ## Choose the variable to display
        legend = TRUE,
        col.regions = quantPal)

## Qualitative variable, whichMax
mapView(sfMapVotes0,
        zcol = "whichMax",
        legend = TRUE,
        col.regions = qualPal)
```

---

[6]In previous sections the spatial object included a modification to the original shapefile in order to display the Canarian Islands in the right bottom corner of the maps. This modification is not needed with `mapview`, so `st_read` imports the shapefile `spMapVotes0` (Section 15.2).

FIGURE 9.9: Snapshot of the interactive quantitative choropleth map produced with `mapview`.

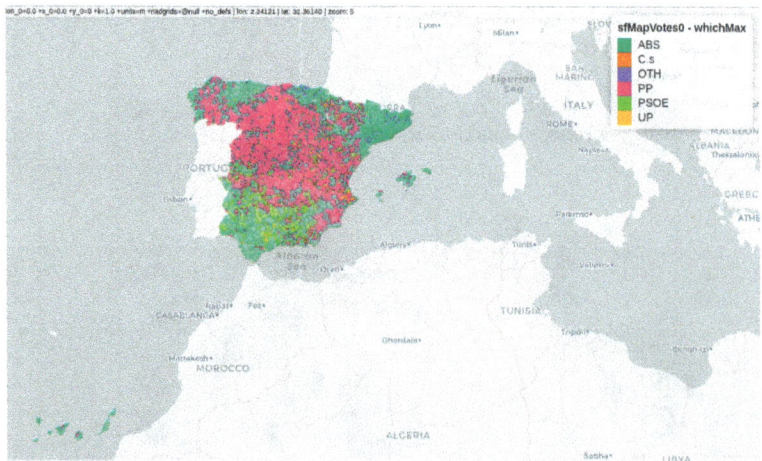

FIGURE 9.10: Snapshot of the interactive qualitative choropleth map produced with `mapview`.

Chapter 10

Thematic Maps: Cartogram Maps

A cartogram is a thematic map that purposely distorts spatial relationships based on a variable. This chapter works with three forms of cartograms: contiguous area cartograms, where the polygons are scaled and distorted while maintaining adjacent edges; non-contiguous area cartograms, which modify the area of the polygon while preserving the shape; and Dorling cartograms, which represent the polygons with non-overlapping circles.

This chapter presents examples of these forms of cartograms to display the gross domestic product and population variables.

The most relevant packages used in this chapter are: sf and sf for reading spatial data; and cartogram for computing the transformations; ggplot2 for creating the maps.

DOI: 10.1201/9781003485384-10

10.1 Introduction

The graphics in this chapter display the gross domestic product (GDP) and population of each province in Spain with three different types of cartograms.

The package `cartogram` implements this type of visual representation. It is based on `sf`. Thus, the graphics in this chapter will be produced with `ggplot2`:

```r
library("ggplot2")
library("sf")
library("cartogram")
```

10.2 Data

The data is available in the shapefile `sfPopGDPSpain` (read Chapter 15 for details). Figure 10.1 contains three scatterplots: GDP for each province, population for each province, and GDP versus population. It is clear that two provinces, Madrid and Barcelona, have a population and a GDP significantly higher than the rest. Besides, the relation between these variables is approximately linear.

```r
sfPopGDPSpain <- st_read("data/Spatial/sfPopGDPSpain.shp")

ggplot(sfPopGDPSpain) +
  geom_point(aes(y = Province, x = Population/1e6)) +
  xlab("Population (million)") +
  theme_bw()

ggplot(sfPopGDPSpain) +
  geom_point(aes(y = Province, x = GDP/1e6)) +
  xlab("GDP (million euros)") +
  theme_bw()

ggplot(sfPopGDPSpain, aes(y = Population/1e6, x = GDP/1e6)) +
  geom_point() +
  geom_smooth() +
  xlab("GDP (million euros)") + ylab("Population (million)") +
  theme_bw()
```

(a) Population

(b) GDP

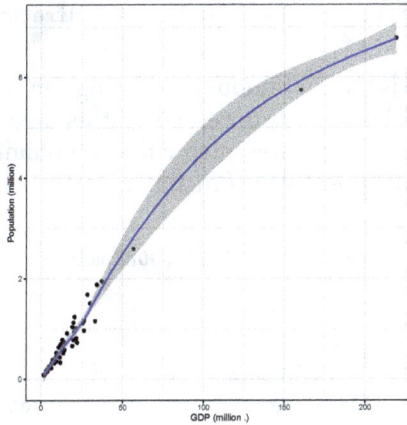

(c) GDP vs Population.

FIGURE 10.1: Population and GDP of the provinces in Spain (2020)

10.3 Cartograms

The first method is the non-contiguous area cartogram. This solution enlarges or reduces each polygon in size according to the variable while preserving its shape.

The function `cartogram_ncont` computes the transformation based on the variable identified by the argument `weight`, and returns a `sf` object. The result is displayed with `geom_sf` (Figure 10.2).

```
cartPopNCont <- cartogram_ncont(sfPopGDPSpain, weight = "
    Population")

ggplot(cartPopNCont) +
  geom_sf(aes(fill = Population)) +
  scale_fill_distiller(palette = "Blues", direction = 1) +
  theme_bw()

cartGDPNCont <- cartogram_ncont(sfPopGDPSpain, weight = "GDP")

ggplot(cartGDPNCont) +
  geom_sf(aes(fill = GDP)) +
  scale_fill_distiller(palette = "Blues", direction = 1) +
  theme_bw()
```

The second type is the continuous area cartogram. The polygon of each province is both scaled and distorted so that its area is directly proportional to the variable being represented while maintaining adjacent edges.

The `sf` objects produced by the function `cartogram_cont` are displayed in Figure 10.3.

```
cartPopCont <- cartogram_cont(sfPopGDPSpain,
                        weight = "Population")

ggplot(cartPopCont) +
  geom_sf(aes(fill = Population)) +
  scale_fill_distiller(palette = "Blues", direction = 1) +
  theme_bw()

cartGDPCont <- cartogram_cont(sfPopGDPSpain,
                        weight = "GDP")

ggplot(cartPopCont) +
  geom_sf(aes(fill = GDP)) +
  scale_fill_distiller(palette = "Blues", direction = 1) +
  theme_bw()
```

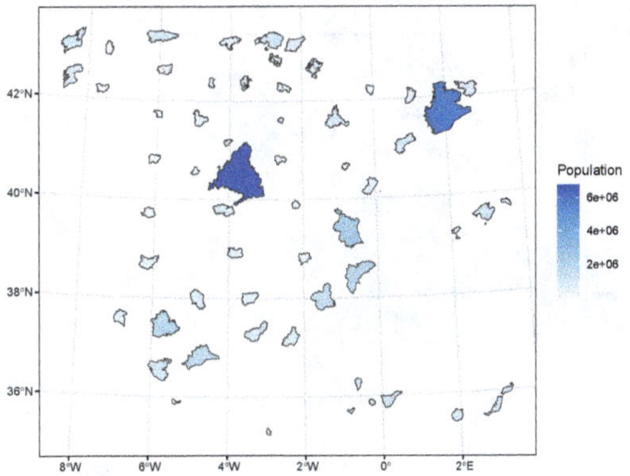

(a) Population

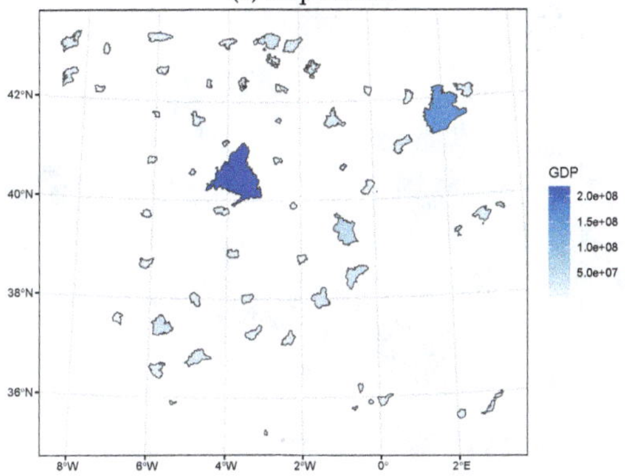

(b) GDP

FIGURE 10.2: Non-contiguous cartograms of population and GDP of the provinces in Spain (2020)

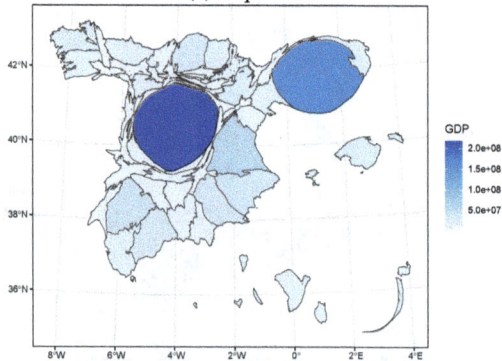

(a) Population

(b) GDP

FIGURE 10.3: Contiguous cartograms of population and GDP of the provinces in Spain (2020)

Finally, the third method is the Dorling cartogram. Provinces are represented by non-overlapping circles, sized proportionally to the data, resembling proportional symbol maps (Chapter 8).

The function `cartogram_dorling` creates the circles, and the result is represented in Figure 10.4.

```
cartPopDorl <- cartogram_dorling(sfPopGDPSpain, weight = "
    Population")

ggplot(cartPopDorl) +
  geom_sf(aes(fill = Population)) +
  scale_fill_distiller(palette = "Blues", direction = 1) +
  theme_bw()

cartGDPDorl <- cartogram_dorling(sfPopGDPSpain, weight = "GDP")

ggplot(cartGDPDorl) +
  geom_sf(aes(fill = GDP)) +
  scale_fill_distiller(palette = "Blues", direction = 1) +
  theme_bw()
```

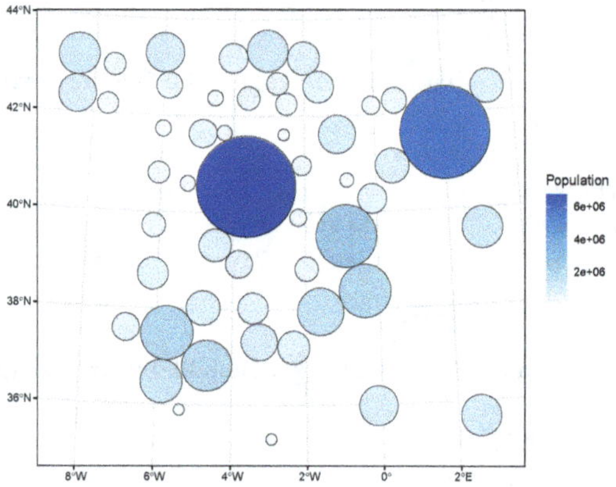

(a) Population

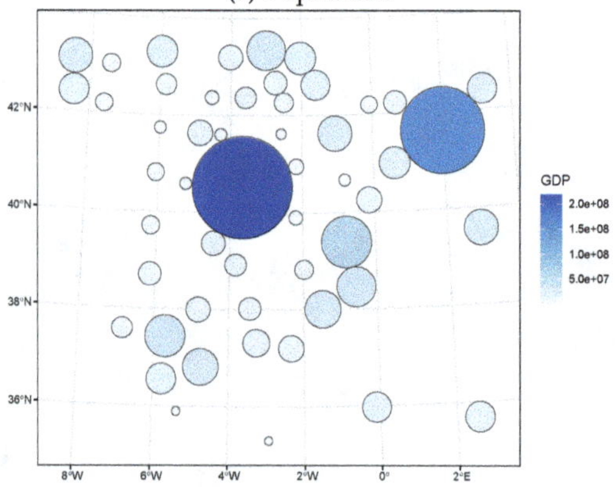

(b) GDP

FIGURE 10.4: Non-overlapping circles cartograms of population and GDP of the provinces in Spain (2020).

Chapter 11

Thematic Maps: Raster Maps

A raster data structure is a matrix of cells organized into rows and columns where each cell contains a value representing information, such as temperature, altitude, population density, land use, etc. This chapter describes how to display a raster with two different data sets: CM-SAF solar irradiation rasters will illustrate the use of quantitative data, and land cover and population data from the NEO-NASA project will exemplify the display of categorical data and multivariate rasters.

Main subjects covered in this chapter are: sequential, qualitative, and diverging palettes; class intervals and visual discrimination; hill shading; and 3D and interactive visualization.

The most relevant packages used in this chapter are: raster and terra for reading raster data; sf for reading spatial vector data; rnaturalearth= and geodata for retrieving data; rasterVis and ggplot2 for visualization of raster data; viridisLite and RColor-Brewer for palettes of colors; classInt for computing class intervals; and mapview for interactive maps.

DOI: 10.1201/9781003485384-11

11.1 Quantitative Data

As an example of quantitative data, this section displays the distribution of annual solar irradiation over the Iberian Peninsula using the estimates from CM SAF. Read Chapter 15 for details about this dataset.

```r
library("raster")
library("terra")
library("sf")

library("rnaturalearth")
library("geodata")

library("viridisLite")

library("rasterVis")

SISavr <- raster("data/Spatial/SISav.nc")

SISavt <- rast("data/Spatial/SISav.nc")
```

The raster of annual averages of solar irradiation estimated by CM SAF can be easily displayed with the `levelplot` method of the `rasterVis` package. Figure 11.1 illustrates this raster with marginal graphics to show the column (longitude) and row (latitude) summaries of the raster object. The summary is computed with the function defined by `FUN.margin` (which uses `mean` as the default value).

```r
levelplot(SISavr)
```

Although the solar irradiation distribution reveals the physical structure of the region, it is recommended to add the geographic context with a layer of administrative boundaries. These boundaries can be retrieved from the Natural Earth service with the rnaturalearth package.

```r
boundarySF <- ne_countries(country = "spain", scale = 50)
## Crop to the limits of the raster object
boundarySF <- st_crop(boundarySF,
                xmin = xmin(SISavt), ymin = ymin(SISavt),
                xmax = xmax(SISavt), ymax = ymax(SISavt))
```

These boundaries are superimposed on the main map with the `layer` and `sp.lines` functions in the `lattice` system (Figure 11.2), and with `geom_sf` in the `ggplot2` system.

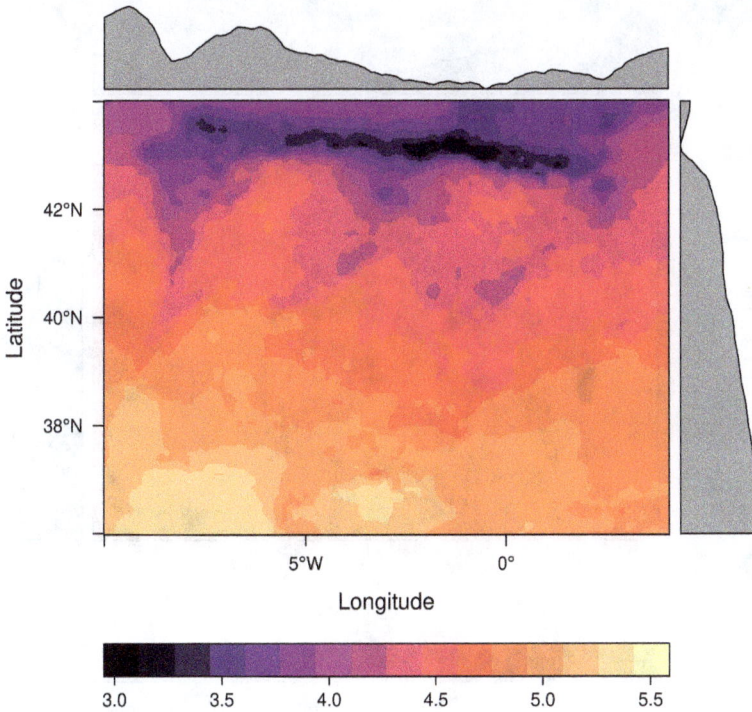

FIGURE 11.1: Annual average of solar radiation displayed with a sequential palette.

```
##ggplot2 version
gplot(SISavt) +
  geom_sf(data = boundarySF, fill = "transparent")

## lattice version
## Convert the sf object to sp
boundarySP <- as(boundarySF, "Spatial")

## Display the data ...
levelplot(SISavt) +
  ## ... and overlay the SpatialLines object
  layer(sp.lines(boundarySP,
            lwd = 0.5))
```

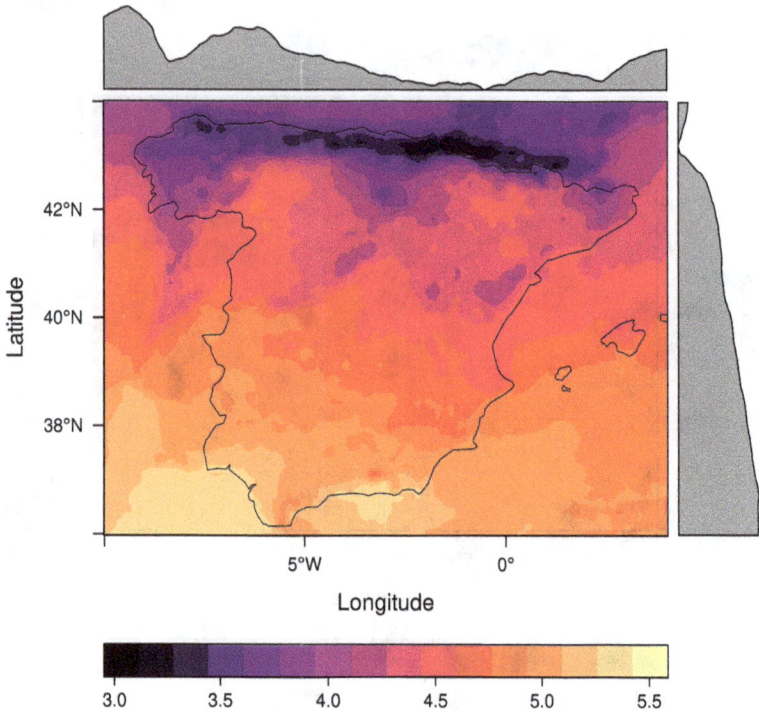

FIGURE 11.2: Annual average of solar radiation with administrative boundaries.

11.1.1 Hill Shading

A frequent method to improve the display of meteorological rasters is the hill shading or shaded relief technique, a method of representing relief on a map by depicting the shadows that would be cast by high ground if light came from a certain sun position (Figure 11.3). The hill shade layer can be computed from the slope and aspect layers derived from a Digital Elevation Model (DEM). This layer will underlay the DEM raster, which will be printed using semitransparency.

The procedure is as follows:

- Download a Digital Elevation Model (DEM) from the DIVA-GIS service with the `geodata` package, obtaining a `SpatRaster` object.

```
DEM <- elevation_30s("ESP", path = tempdir())
```

- Compute the hill shade raster with `terrain` and `shade` from the terra package, or with `terrain` and `hillShade` from raster package, with a certain angle and direction.

```
slope <- terrain(DEM, "slope", unit = "radians")
aspect <- terrain(DEM, "aspect", unit = "radians")
hs <- shade(slope = slope, aspect = aspect,
            angle = 60, direction = 45)

DEMr <- raster(DEM)
sloper <- terrain(DEMr, "slope")
aspectr <- terrain(DEMr, "aspect")
hsr <- hillShade(slope = sloper, aspect = aspectr,
                 angle = 60, direction = 45)
```

- Combine the result with the previous map using semitransparency.

```
## hillShade theme: gray colors and semitransparency
hsTheme <- GrTheme(regions = list(alpha = 0.5))

levelplot(SISavt,
          par.settings = YlOrRdTheme,
          margin = FALSE, colorkey = FALSE) +
  ## Overlay the hill shade raster
  levelplot(hs, par.settings = hsTheme, maxpixels = 1e6) +
  ## and the countries boundaries
  layer(sp.lines(boundarySP, lwd = 0.5))
```

11.1.2 Diverging Palettes

Next, instead of displaying the absolute values of each cell, we will analyze the differences between each cell and the global average value. This average is computed with the `global` function in the terra package and with `cellStats` in the raster package and subtracted from the original raster.

```
meanRad <- global(SISavt, "mean")
meanRad <- as.numeric(meanRad)
SISavt <- SISavt - meanRad
```

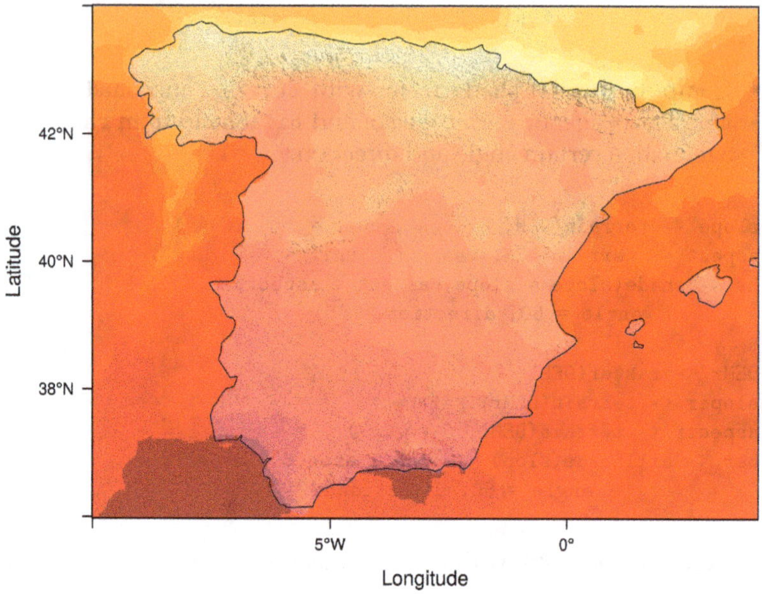

FIGURE 11.3: Hill shading of annual average of solar radiation.

```
meanRad <- cellStats(SISavr, "mean")
SISavr <- SISavr - meanRad
```

Figure 11.4 displays the relation between these scaled values and latitude (y), with five different groups defined by the longitude (cut(x, 5)), using the magma palette contained in the viridisLite package. It is evident that larger irradiation values are associated with lower latitudes. However, there is no such clear relation between irradiation and longitude.

```
xyplot(layer ~ y, data = SISavt,
       groups = cut(x, 5),
       par.settings = rasterTheme(symbol = magma(n = 5,
                                        begin = 0, end = 0.9,
                                        direction = -1)),
       xlab = "Latitude", ylab = "Solar radiation (scaled)",
       auto.key = list(space = "right",
                       title = "Longitude",
                       cex.title = 1.3))
```

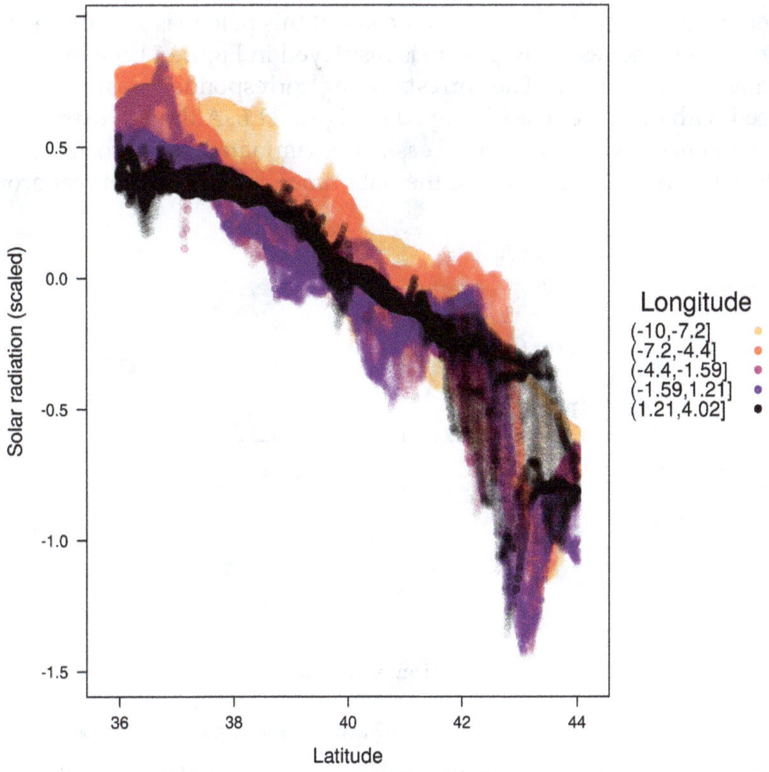

FIGURE 11.4: Relation between scaled annual average radiation and latitude for several longitude groups.

Numerical information ranging in an interval including a neutral value is commonly displayed with diverging palettes. These palettes represent neutral classes with light colors, while low and high extremes of the data range are highlighted using dark colors with contrasting hues. I use the Purple-Orange palette from ColorBrewer with purple for positive values and orange for negative values. In order to underline the position of the interval containing zero, the center color of this palette is substituted with pure white. The resulting palette is displayed in Figure 11.5 with the custom `showPal` function. The corresponding correspondent raster map produced with this palette is displayed in Figure 11.6. Although extreme positive and negative values can be easily discriminated, the zero value is not associated with white because the data range is not symmetrical around zero.

```
divPal <- brewer.pal(n = 9, "PuOr")
divPal[5] <- "#FFFFFF"

showPal <- function(pal)
{
    N <- length(pal)
    image(1:N, 1, as.matrix(1:N), col = pal,
        xlab = "", ylab = "",
        xaxt = "n", yaxt = "n",
        bty = "n")
}

showPal(divPal)

divTheme <- rasterTheme(region = divPal)

levelplot(SISavt, contour = TRUE, par.settings = divTheme)
```

The solution is to connect the symmetrical color palette with the asymmetrical data range. The first step is to create a set of breaks such that the zero value is the center of one of the intervals.

```
rng <- range(SISavt[])
## Number of desired intervals
nInt <- 15
## Increment corresponding to the range and nInt
inc0 <- diff(rng)/nInt
## Number of intervals from the negative extreme to zero
n0 <- floor(abs(rng[1])/inc0)
```

FIGURE 11.5: Purple-Orange diverging palette using white as the middle color.

```
## Update the increment adding 1/2 to position zero in the center
      of an interval
inc <- abs(rng[1])/(n0 + 1/2)
## Number of intervals from zero to the positive extreme
n1 <- ceiling((rng[2]/inc - 1/2) + 1)
## Collection of breaks
breaks <- seq(rng[1], by = inc, length= n0 + 1 + n1)
```

The next step is to compute the midpoints of each interval. These points represent the data belonging to each interval, and their value will be connected with a color of the palette.

```
## Midpoints computed with the median of each interval
idx <- findInterval(SISavt[], breaks, rightmost.closed = TRUE)
mids <- tapply(SISavt[], idx, median)
## Maximum of the absolute value both limits
mx <- max(abs(breaks))
```

A simple method to relate the palette and the intervals is with a straight line such that a point is defined by the absolute maximum value, ((mx, 1)), and another point by zero, ((0, 0.5)). Why are we using the interval [0, 1] as the y-coordinate of this line, and why is 0.5 the result of zero? The reason is that the input of the break2pal function will be the result of colorRamp, a function that creates another interpolating function that maps colors with values between 0 and 1. Therefore, a new palette is created, extracting colors from the original palette, such that the central color

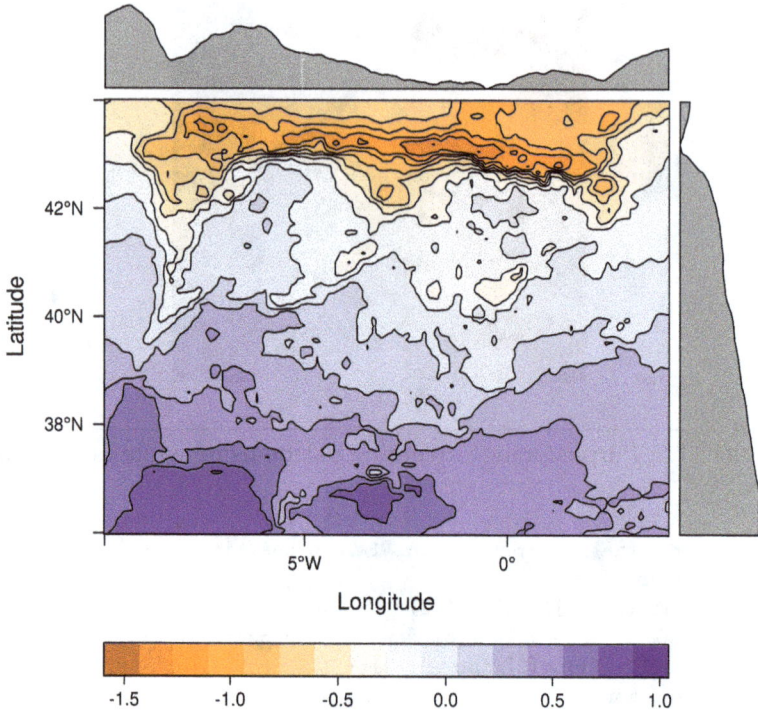

FIGURE 11.6: Asymmetric raster data (scaled annual average irradiation) displayed with a symmetric diverging palette.

(white) is associated with the interval containing zero. This palette is displayed in Figure 11.7.

The raster map produced with this new palette is displayed in Figure 11.8. Now zero is clearly associated with the white color.

```
break2pal <- function(x, mx, pal){
    ## x = mx gives y = 1
    ## x = 0 gives y = 0.5
    y <- 1/2*(x/mx + 1)
    rgb(pal(y), maxColorValue = 255)
}

## Interpolating function that maps colors with [0, 1]
## rgb(divRamp(0.5), maxColorValue=255) gives "#FFFFFF" (white)
```

FIGURE 11.7: Modified diverging palette related to the asymmetrical raster data.

```
divRamp <- colorRamp(divPal)
## Diverging palette where white is associated with the interval
## containing the zero
pal <- break2pal(mids, mx, divRamp)
showPal(pal)

levelplot(SISavt,
          par.settings = rasterTheme(region = pal),
          at = breaks, contour = TRUE)
```

It is interesting to note two operations carried out internally by the lattice package. First, the custom.theme function (used by rasterTheme) creates a new palette with 100 colors using colorRampPalette to interpolate the palette passed as an argument. Second, the level.colors function makes the arrangement between intervals and colors. If this function receives more colors than intervals, it chooses a subset of the palette, disregarding some of the intermediate colors. Therefore, because this function will receive 100 colors from par.settings, it is difficult to control exactly which colors of our original palette will be represented.

An alternative way for finer control is to fill the regions$col component of the theme with our palette after it has been created (Figure 11.9).

```
divTheme <- rasterTheme(regions = list(col = pal))

levelplot(SISavt,
```

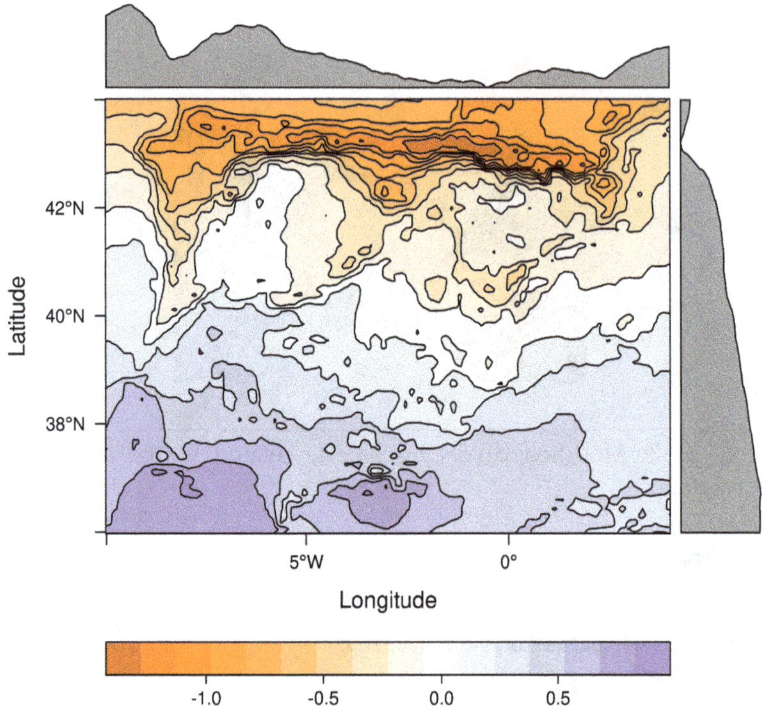

FIGURE 11.8: Asymmetric raster data (scaled annual average irradiation) displayed with a modified diverging palette.

```
par.settings = divTheme,
at = breaks,
contour = TRUE)
```

A final improvement to this map is to compute the intervals using a classification algorithm with the `classInt` package. With this approach it is likely that zero will not be perfectly centered in its corresponding interval. The remaining code is exactly the same as above, replacing the `breaks` vector with the result of the `classIntervals` function. Figure 11.10 displays the result.

```
cl <- classIntervals(SISavt[], style = "kmeans")
breaks <- cl$brks
```

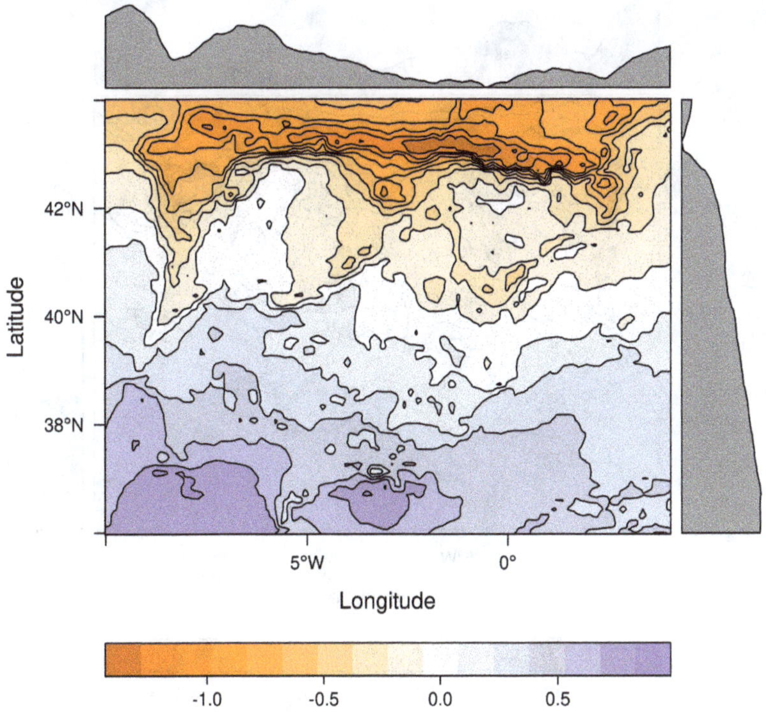

FIGURE 11.9: Same as Figure 11.8 but colors are assigned directly to the regions$col component of the theme.

```
## Repeat the procedure previously exposed, using the 'breaks'
    vector
## computed with classIntervals
idx <- findInterval(SISavt[], breaks, rightmost.closed = TRUE)
mids <- tapply(SISavt[], idx, median)

mx <- max(abs(breaks))
pal <- break2pal(mids, mx, divRamp)

## Modify the vector of colors in the 'divTheme' object
divTheme$regions$col <- pal

levelplot(SISavt,
          par.settings = divTheme,
```

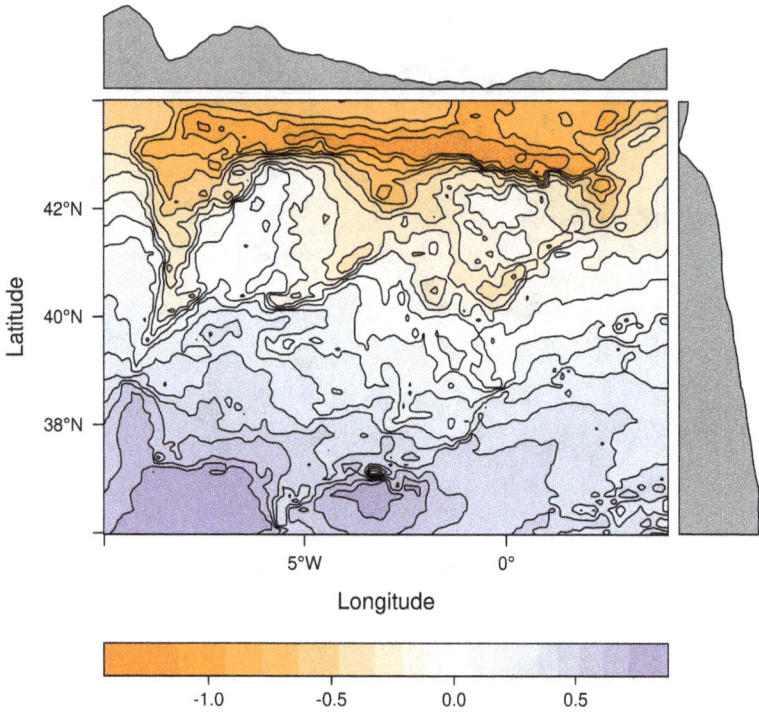

FIGURE 11.10: Same as Figure 11.9 but defining intervals with the optimal classification method.

```
at = breaks,
contour = TRUE)
```

11.2 Categorical Data

Land cover is the observed physical cover on the Earth's surface. A set of seventeen different categories is commonly used. Using satellite observations, it is possible to map where on Earth each of these seventeen land surface categories can be found and how these land covers change over time.

This section illustrates how to read and display rasters with categorical information using information from the NEO-NASA project. Read Chapter 15 for details about these datasets.

The main packages in this section are: `raster` and `terra` are used for reading and manipulating the raster data, and `rasterVis` for producing the graphics.

After the land cover and population density files have been downloaded, two `RasterLayer` objects can be created with the `raster` package or two `SpatRaster` objects with the `terra` package. Both files are read, their geographical extent reduced to the area of India and China, and cleaned (99999 cells are replaced with `NA`).

```
## raster
myExtR <- extent(65, 135, 5, 55)

popR <- raster("data/Spatial/875430rgb-167772161.0.FLOAT.TIFF")
popR <- crop(popR, myExtR)
popR[popR==99999] <- NA

landClassR <- raster("data/Spatial/241243rgb-167772161.0.TIFF")
landClassR <- crop(landClassR, myExtR)

## terra
myExtT <- ext(65, 135, 5, 55)

popT <- rast("data/Spatial/875430rgb-167772161.0.FLOAT.TIFF")
names(popT) <- "population"
popT <- crop(popT, myExtT)
popT[popT==99999] <- NA

landClassT <- rast("data/Spatial/241243rgb-167772161.0.TIFF")
names(landClassT) <- "landClass"
landClassT <- crop(landClassT, myExtT)
```

Each land cover type is designated with a different key: the sea is labeled with 0; forests with 1 to 5; shrublands, grasslands, and wetlands with 6 to 11; agriculture and urban lands with 12 to 14; and snow and barren with 15 and 16. These four groups (sea is replaced by `NA`) will be the levels of the categorical raster. The `raster` package includes the `ratify` method to define a layer as categorical data, filling it with integer values associated to a Raster Attribute Table (RAT).

```
landClassR[landClassR %in% c(0, 254)] <- NA
## Only four groups are needed:
## Forests: 1:5
## Shrublands, etc: 6:11
## Agricultural/Urban: 12:14
```

```
## Snow: 15:16
landClassR <- cut(landClassR, c(0, 5, 11, 14, 16))
## Add a Raster Attribute Table and define the raster as
    categorical data
landClassR <- ratify(landClassR)
## Configure the RAT: first create a RAT data.frame using the
## levels method; second, set the values for each class (to be
## used by levelplot); third, assign this RAT to the raster
## using again levels
rat <- levels(landClassR)[[1]]
rat$classes <- c("Forest", "Land", "Urban", "Snow")
levels(landClassR) <- rat

landClassT[landClassT %in% c(0, 254)] <- NA
landClassT <- classify(landClassT, c(0, 5, 11, 14, 16))

rat <- levels(landClassT)[[1]]
names(rat) <- c("ID", "classes")
rat$classes <- c("Forest", "Land", "Urban", "Snow")
levels(landClassT) <- rat
```

This categorical raster can be displayed with the `levelplot` method of the `rasterVis` package. Previously, a theme was defined with the background color set to `lightskyblue1` to display the sea areas (filled with `NA` values), and the region palette was defined with adequate colors (Figure 11.11).

```
qualPal <- c("palegreen4", # Forest
             "lightgoldenrod", # Land
             "indianred4", # Urban
             "snow3") # Snow

qualTheme <- rasterTheme(region = qualPal,
                         panel.background = list(col = "lightskyblue1"
                         )
             )

levelplot(landClassT, maxpixels = 3.5e5,
          par.settings = qualTheme)
```

Let's explore the relation between the land cover and population density rasters. Figure 11.12 displays this latter raster using a logarithmic scale, defined with `zscaleLog`.

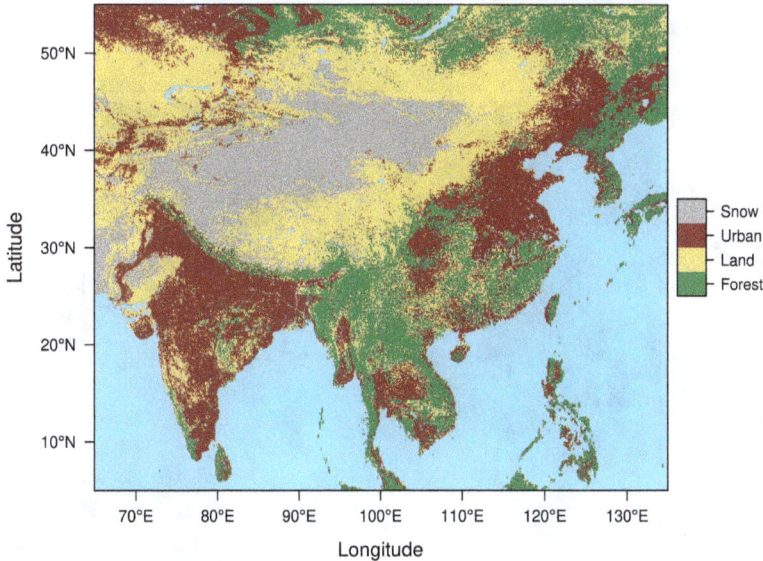

FIGURE 11.11: Land cover raster (categorical data).

```
pPop <- levelplot(popT, zscaleLog = 10,
              par.settings = BTCTheme,
              maxpixels = 3.5e5)
pPop
```

Both rasters can be joined together with the stack method to create a new RasterStack object or with c to create a multilayer SpatStat object. Figure 11.13 displays the distribution of the logarithm of the population density associated with each land class.

```
## Join the RasterLayer objects to create a RasterStack object.
s <- stack(popR, landClassR)
names(s) <- c("pop", "landClass")

## Join the SpatRaster objects to create a multilayer object.
st <- c(popT, landClassT)
names(st) <- c("pop", "landClass")

densityplot(~log10(pop), ## Represent the population
        groups = landClass, ## grouping by land classes
        data = s,
```

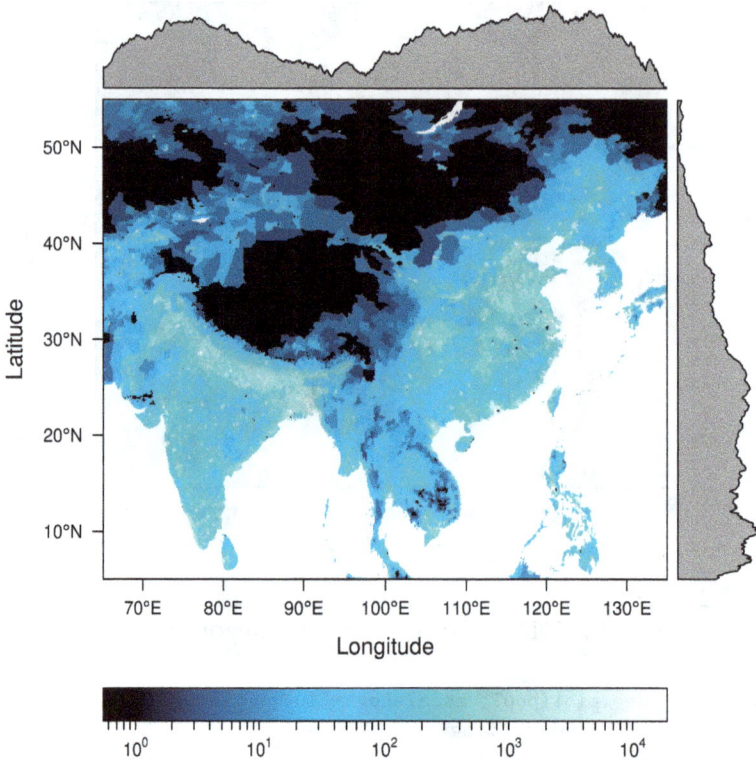

FIGURE 11.12: Population density raster.

```
## Do not plot points below the curves
plot.points = FALSE)
```

11.3 ✺Bivariate Legend

We can reproduce the code used to create the multivariate choropleth (Section 9.5) using the `levelplot` function from the `rasterVis` package. Again, the result is a list of `trellis` objects. Each of these objects is the representation of the population density in a particular land class.

```
classes <- rat$classes
nClasses <- length(classes)
```

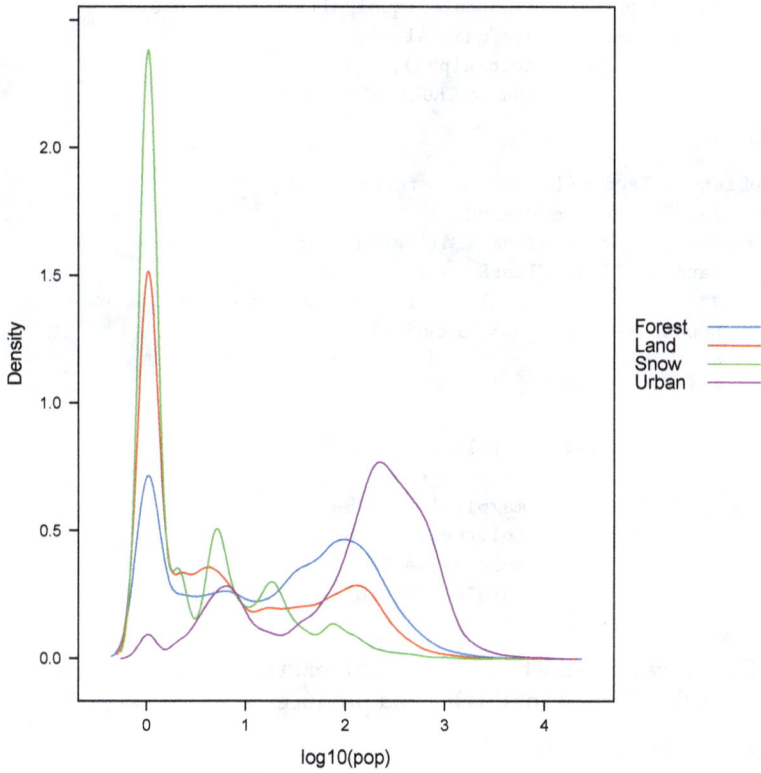

FIGURE 11.13: Distribution of the logarithm of the population density associated with each land class.

```
logPopAt <- c(0, 0.5, 1.85, 4)

nIntervals <- length(logPopAt) - 1

multiPal <- sapply(1:nClasses, function(i)
{
    colorAlpha <- adjustcolor(qualPal[i], alpha = 0.4)
    colorRampPalette(c(qualPal[i],
                       colorAlpha),
                  alpha = TRUE)(nIntervals)
})

pList <- lapply(1:nClasses, function(i){
    landSub <- landClassR
    ## Those cells from a different land class are set to NA...
    landSub[!(landClassR == i)] <- NA
    ## ... and the resulting raster masks the population raster
    popSub <- mask(popR, landSub)
    ## Palette
    pal <- multiPal[, i]

    pClass <- levelplot(log10(popSub),
                        at = logPopAt,
                        maxpixels = 3.5e5,
                        col.regions = pal,
                        colorkey = FALSE,
                        margin = FALSE)
})
```

The +.trellis function of the latticeExtra package with Reduce superposes the elements of this list and produces a trellis object.

```
p <- Reduce('+', pList)
```

The legend is created with grid.raster and grid.text, following the same procedure exposed in section 9.5.

```
library("grid")

legend <- layer(
{
    ## Center of the legend (rectangle)
    x0 <- 125
    y0 <- 22
    ## Width and height of the legend
```

```r
w <- 10
h <- w / nClasses * nIntervals
## Legend
grid.raster(multiPal, interpolate = FALSE,
              x = unit(x0, "native"),
              y = unit(y0, "native"),
          width = unit(w, "native"))
## Axes of the legend
## x-axis (qualitative variable)
grid.text(classes,
        x = unit(seq(x0 - w * (nClasses -1)/(2*nClasses),
                  x0 + w * (nClasses -1)/(2*nClasses),
                  length = nClasses),
              "native"),
        y = unit(y0 + h/2, "native"),
        just = "bottom",
        rot = 10,
        gp = gpar(fontsize = 6))
## y-axis (quantitative variable)
yLabs <- paste0("[",
              paste(logPopAt[-nIntervals],
                  logPopAt[-1], sep = ","),
              "]")
grid.text(yLabs,
        x = unit(x0 + w/2, "native"),
        y = unit(seq(y0 - h * (nIntervals -1)/(2*nIntervals),
                  y0 + h * (nIntervals -1)/(2*nIntervals),
                  length = nIntervals),
              "native"),
        just = "left",
        gp = gpar(fontsize = 6))

})
```

Figure 11.14 displays the result with the legend.

```r
p + legend
```

11.4 Interactive Graphics

11.4.1 3D Visualization

An alternative method for a DEM is 3D visualization, where the user can rotate or zoom the figure. This solution is available thanks to the

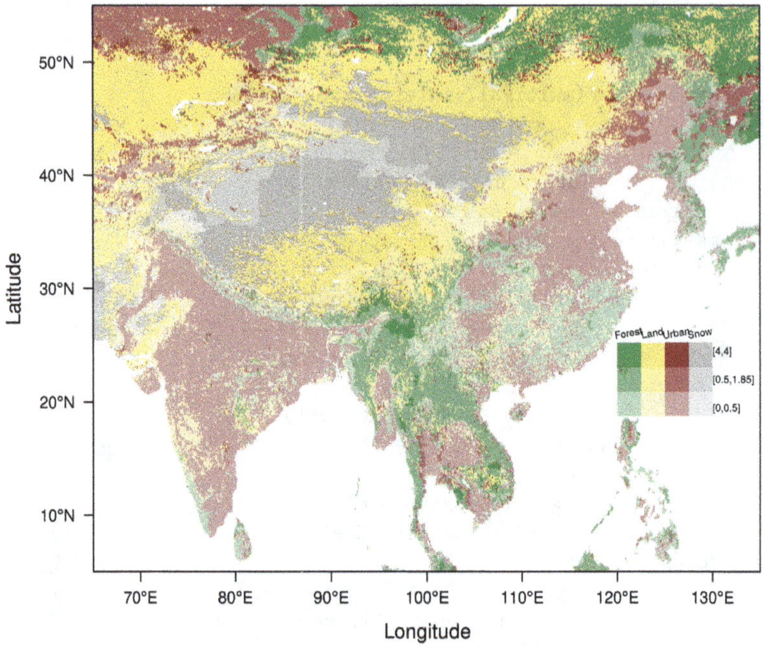

FIGURE 11.14: Population density for each land class (multivariate raster).

rgl package, which provides functions for 3D interactive graphics. The plot3D function in the rasterVis package is a wrapper to this package for RasterLayer objects (Figure 11.15).

```
plot3D(DEMr, maxpixels = 5e4)
```

The output scene can be exported to several formats, such as STL with writeSTL, a format commonly used in 3D printing.

```
library("rgl")

writeSTL("docs/images/rgl/DEM.stl")
```

11.4.2 Mapview

The package mapview is able to work with Raster* objects. Thus, the SISavr object can be easily displayed in an interactive map with the next

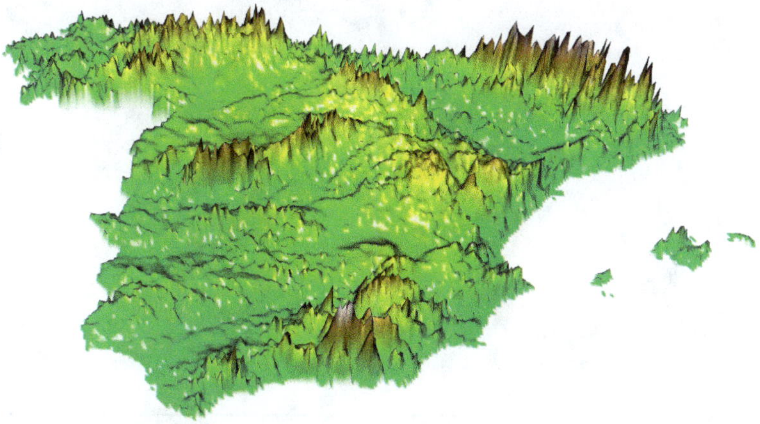

FIGURE 11.15: 3D visualization of a Digital Elevation Model.

code. However, it must be noted that, unlike with vector data (Sections 8.6.1 and 9.6), the interactivity of this map is restricted to zoom and movement. The mouse hovering or clicking does not produce any result.

```
library("mapview")

mvSIS <- mapview(SISavr, legend = TRUE)
```

This map can be improved with another layer of information, the name and location of the meteorological stations of the SIAR network. This information is stored in the file SIAR.csv.

```
SIAR <- read.csv("data/Spatial/SIAR.csv")
```

These points can be encoded as a SpatialPointsDataFrame object with the sp package:

```
spSIAR <- SpatialPointsDataFrame(coords = SIAR[, c("lon", "lat")
    ],
                                 data = SIAR,
                                 proj4str = CRS(projection(SISavr)))
```

or a sf object:

```
sfSIAR <- st_as_sf(SIAR,
                   coords = c("lon", "lat"),
                   crs = crs(SISavt))
```

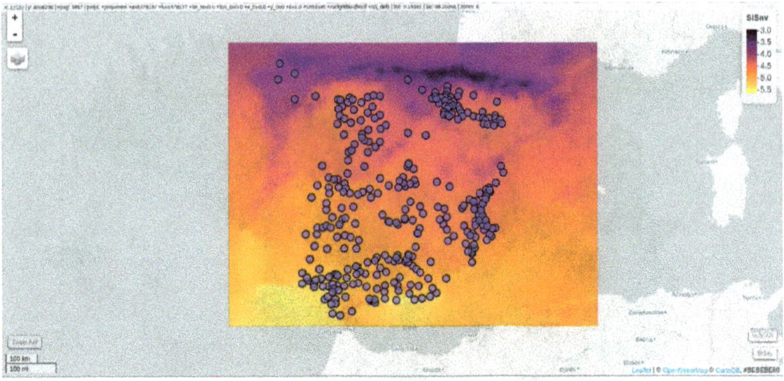

FIGURE 11.16: Snapshot of the interactive map produced with `mapview` combining a `RasterLayer` and a `SpatialPointsDataFrame` objects.

Both objects, as shown in section 8.6.1, can be displayed with `mapview`. The resulting map is reactive to mouse hovering and clicking.

```
mvSIAR <- mapview(sfSIAR,
                  label = sfSIAR$Estacion)
```

Both layers of information can be combined with the + operator. Figure 11.16 shows a snapshot of this interactive map.

```
mvSIS + mvSIAR
```

Chapter 12

Vector Fields

Many objects in our natural environment exhibit directional features that are naturally represented by vector data. Vector fields, commonly found in science and engineering, describe the spatial distribution of a vector variable such as fluid flow or electromagnetic forces. A suitable visualization method has to display both the magnitude and the direction of the vectors at any point.

This chapter illustrates two visualization techniques, arrow plots and streamlines, included in the rasterVis package.

DOI: 10.1201/9781003485384-12

12.1 Introduction

The graphics of this chapter display the vector fields defined by the wind direction and speed forecast published in the THREDSS server[1] of Meteogalicia[2]. This server provides access through different protocols to the output of a Weather Research and Forecasting (WRF) model, a mesoscale numerical weather prediction system.

This vector field is encoded in a `RasterStack` object with two layers: wind speed or vector magnitude and wind direction or vector direction.

```
library("raster")
library("rasterVis")
library("RColorBrewer")

## Local vector direction
wDir <- raster('data/Spatial/wDir')/180*pi
## Local vector magnitude
wSpeed <- raster('data/Spatial/wSpeed')
## Vector field encoded in a RasterStack with two layers,
    magnitude
## and direction
windField <- stack(wSpeed, wDir)
names(windField) <- c('magnitude', 'direction')
```

12.2 Arrow Plot

A frequent vector visualization technique is the arrow plot, which draws a small arrow at discrete points within the vector field (Figure 12.1). This approach is best suited for small datasets. If the grid of discrete points gets too dense or if the variations in magnitude are too big, the images tend to be visually confusing.

The `rasterVis` package includes the function `vectorplot`, based on `lattice`[3]. This function is able to display vector fields computed from a `Raster` object. Moreover, as the next example illustrates, this function displays `RasterStack` and `RasterBrick` objects with two layers encoding a vector field, slope (local vector magnitude), and aspect (local vector direction).

---

[1]http://mandeo.meteogalicia.es/thredds/catalogos/WRF_2D/catalog.html
[2]http://www.meteogalicia.es
[3]A `ggplot2` solution is available in the `ggquiver` package (https://github.com/mitchelloharawild/ggquiver), although it does not understand `Raster` objects.

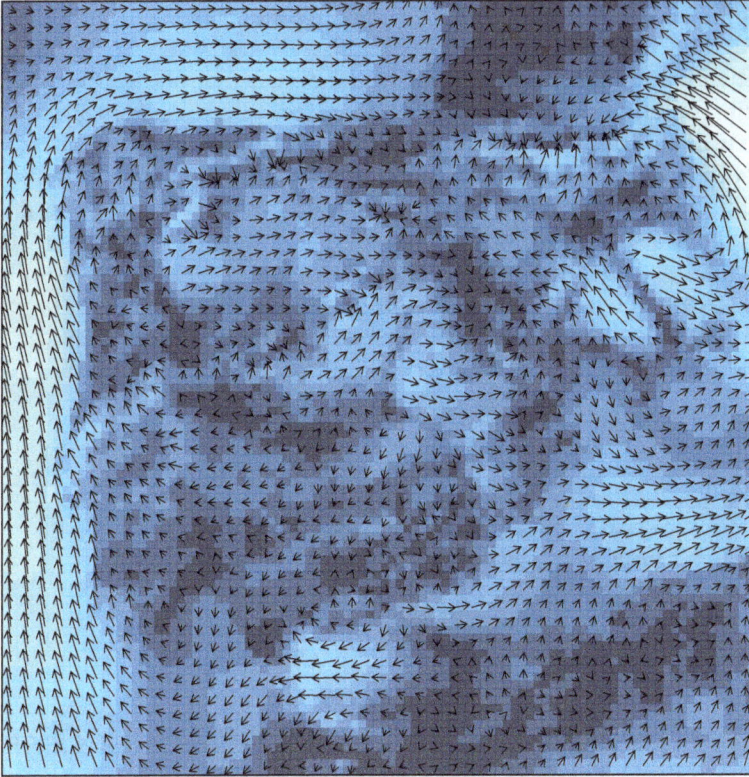

FIGURE 12.1: Arrow plot of the wind vector field.

```
vectorTheme <- BTCTheme(regions = list(alpha = 0.7))

vectorplot(windField,
           isField = TRUE, ##RasterStack is a vector field
           aspX = 5, aspY = 5, ##Multipliers to adjust the relation
                               ##between slope/aspect and
                               ##horizontal/vertical displacements in
                               ##the figure.
           scaleSlope = FALSE, ## Slope values are *not* scaled
           par.settings = vectorTheme,
           colorkey = FALSE,
           scales = list(draw = FALSE))
```

12.3 Streamlines

Another solution is to depict the directional structure of the vector field by its integral curves, also denoted as flow lines or streamlines. There are a variety of algorithms to produce such visualization. The `streamplot` function of `rasterVis` displays streamlines with a procedure inspired by the FROLIC algorithm (Wegenkittl and Gröller 1997): For each point, *droplet*, of a jittered regular grid, a short streamline portion, *streamlet*, is calculated by integrating the underlying vector field at that point. The main color of each streamlet indicates local vector magnitude. Streamlets are composed of points whose sizes, positions, and color degradation encode the local vector direction (Figure 12.2).

```
myTheme <- streamTheme(
    region = rev(brewer.pal(n = 4, "Greys")),
    symbol = rev(brewer.pal(n = 9, "Blues")))

streamplot(windField, isField = TRUE,
        par.settings = myTheme,
        droplet = list(pc = 12), ## Amount of droplets,
            percentage of cells
        streamlet = list(L = 5, ## Length of the streamlet
                        h = 5), ## Calculation step
        scales = list(draw = FALSE),
        panel = panel.levelplot.raster)
```

The magic of Figures 12.1 and 12.2 is that they show the underlying physical structure of the spatial region, only displaying wind speed and direction. It is easy to recognize the Iberian Peninsula surrounded by strong winds along the eastern and northern coasts. Another feature easily distinguishable is the Strait of Gibraltar, a channel that connects the Atlantic Ocean to the Mediterranean Sea between the south of Spain and the north of Morocco. Also apparent are the Pyrenees Mountains and some of the river valleys.

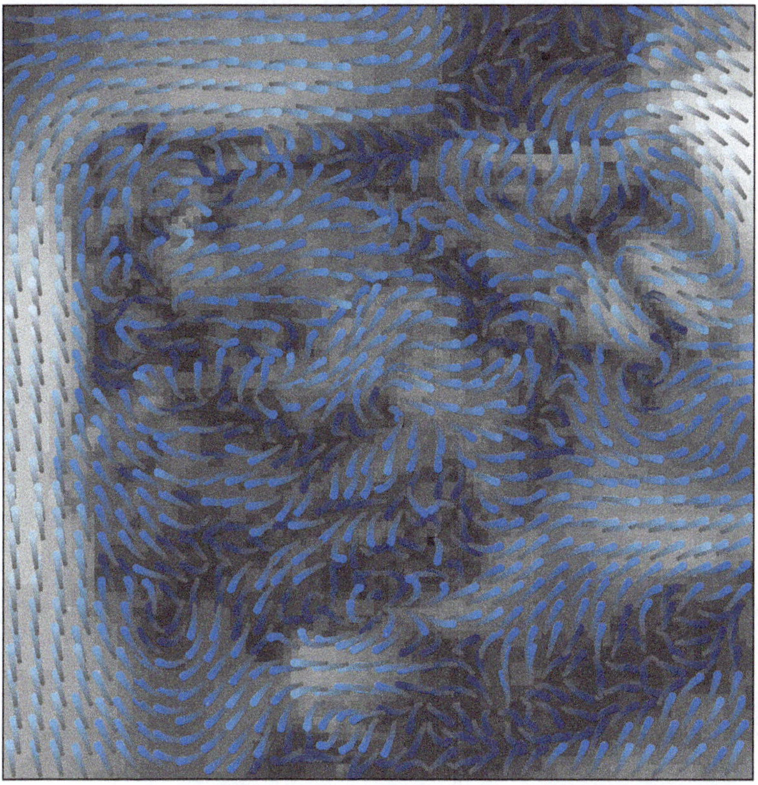

FIGURE 12.2: Streamlines of the wind vector field.

Chapter 13

Physical Maps

A physical map shows the physical landscape features of a place. Mountains and elevation changes are usually shown with different colors and shades to show relief, using green to show lower elevations and browns for high elevations.

This chapter details how to create a physical map of Brazil with data from different sources, covering label positioning and map overlaying.

The most relevant packages used in this chapter are: raster and terra for reading raster data; rasterVis and tidyterra for visualization of raster data; colorspace and RColorBrewer for palettes of colors; and rnaturalearth and geodata for retrieving spatial data.

DOI: 10.1201/9781003485384-13

13.1 Introduction

Brazil[1], the world's fifth largest country, is one of the seventeen megadiverse countries[2], home to diverse wildlife, natural environments, and extensive natural resources in a variety of protected habitats. Throughout this section we will create a physical map of this exceptional country using data from several data services.

13.1.1 Packages

Several packages are employed to produce the map:

- `terra` for reading and manipulating raster data.

- `sp` and `sf` for reading and manipulating spatial vector data.

```
library("terra")

library("sf")
library("sp")
```

- `rasterVis` and `tidyterra` for creating maps with `lattice` and `ggplot2`, respectively.

- `RColorBrewer` and `colorspace` for defining the palette of colors.

- `ggrepel` for text labeling.

```
library("rasterVis")
library("tidyterra")

library("colorspace")

library("ggrepel")
```

- `rnaturalearth` and `geodata` for retrieving data from the NaturalEarth[3] and the GeoData[4] services, respectively.

---

[1]http://en.wikipedia.org/wiki/Brazil
[2]http://en.wikipedia.org/wiki/Megadiverse_countries
[3]http://www.naturalearthdata.com/
[4]https://diva-gis.org/

```r
library("rnaturalearth")
library("rnaturalearthhires")
library("geodata")
```

13.2 Retrieving Data

Three types of information are needed: administrative boundaries, terrain elevation, and rivers and lakes.

1. The administrative boundaries are available from Natural Earth with the packages rnaturalearth and rnaturalearthhires[5].

   ```r
   brazilAdm <- ne_states(country = "brazil")
   ```

2. The terrain elevation or digital elevation model (DEM) is available from the DIVA-GIS service with the geodata package, obtaining a SpatRaster object.

   ```r
   brazilDEM <- elevation_30s("BRA", mask = FALSE,
                              path = tempdir())
   ```

3. The water lines (rivers and lakes) are available from Natural Earth Data.

   ```r
   worldRiv <- ne_download(type = "rivers_lake_centerlines",
                           category = "physical",
                           scale = 10)
   ```

The rivers and lakes database from Natural Earth Data comprises the whole world's extent, but we only need the rivers of Brazil. The function st_crop of the package sf determines the intersection between two objects.

```r
## only those features labeled as "River" are needed
worldRiv <- worldRiv[worldRiv$featurecla == "River",]

## Ensure the CRS of brazilDEM matches the CRS of worldRiv
crs(brazilDEM) <- crs(worldRiv)

## and intersect it with worldRiv to extract brazilian rivers
## from the world database
brazilRiv <- st_crop(worldRiv, brazilDEM)
```

---

[5]The package rnaturalearthhires is not available at CRAN due to its size. More information: https://docs.ropensci.org/rnaturalearthhires/

13.3 Labels

Each region of Brazil will be labeled with the name of its corresponding polygon. The locations of the labels are defined by the `latitude` and `longitude` values included in the `sf` object. In addition, a larger label with the name of the country will be placed in the average centroid.

```
## Locations of labels of each polygon
centroids <- brazilAdm[ , c("longitude", "latitude")]
## Extract the data
centroids <- st_drop_geometry(centroids)
## Location of the "Brazil" label (average of the centroids)
xyBrazil <- apply(centroids, 2, mean)
```

Some region names are too long to be displayed in one line. Thus, a previous step is to split the string if it comprises more than two words.

```
admNames <- strsplit(as.character(brazilAdm$name), ' ')

admNames <- sapply(admNames,
            FUN = function(s){
                sep = if (length(s)>2) '\n' else ' '
                paste(s, collapse = sep)
            })
```

13.4 Overlaying Layers of Information

Therefore, the physical map is composed of three layers: the altitude raster layer, the rivers defined by the `brazilRiv` object, and the administrative boundaries defined by the `brazilAdm` object and their labels (`admNames` and `centroids`).

The altitude raster layer can be displayed with the `levelplot` function of the `rasterVis` package (`lattice` version) or with the `geom_spatraster` function of the `tidyterra` package (ggplot2 version).

With `levelplot`, we use a terrain colors palette, such as the one produced by the `terrain_hcl` function from the `colorspace` package (Ihaka et al. 2024). The theme for the graphic is configured via `rasterTheme` with this palette and setting the background (`NA` values) to a light blue (Figure 13.4).

```
terrainTheme <- rasterTheme(region = terrain_hcl(15),
                    panel.background = list(col = "
                        lightskyblue1"))
```

```
altPlot <- levelplot(brazilDEM, par.settings = terrainTheme,
                    maxpixels = 1e6, panel = panel.levelplot.raster,
                    margin = FALSE, colorkey = FALSE)
```

The tidyterra package includes multiple palettes[6] that are well suited for representation of digital elevation models. For example, the scale_-fill_whitebox_c function encodes palettes from the WhiteBox project[7], of which I have chosen the high_relief one (Figure 13.4).

```
ggplot() +
  geom_spatraster(data = brazilDEM,
                  show.legend = FALSE) +
  scale_fill_whitebox_c("high_relief",
                        na.value = "aquamarine") +
  theme_bw()
```

Under the lattice framework, the rivers can be represented with the sp.lines function of the sp package and superimposed over the DEM map with the layer mechanism of the latticeExtra package. First, brazilRiv must be converted to a SpatialLines object.

```
brazilRivsp <- as_Spatial(brazilRiv)
```

On the ggplot2 side, the geom_sf function of the sf package displays the brazilRiv without a previous conversion.

Then, the brazilAdm object must be converted to a SpatialPolygons object to be displayed with sp.polygons and lattice, while the geom_sf function can use this object directly.

```
brazilAdmsp <- as_Spatial(brazilAdm)
```

Finally, the labels will be printed with panel.text for the lattice version and with the geom_text_repel function of the ggrepel package[8], to repel overlapping text labels for the ggplot2 version.

The final result is shown in Figure 13.2.

```
## lattice version
altPlot + layer({
  ## Rivers
  sp.lines(brazilRivsp, col = 'darkblue', lwd = 0.2)
  ## Administrative boundaries
```

---

[6]https://dieghernan.github.io/tidyterra/articles/palettes.html
[7]https://www.whiteboxgeo.com/
[8]https://ggrepel.slowkow.com/

(a) `lattice` version

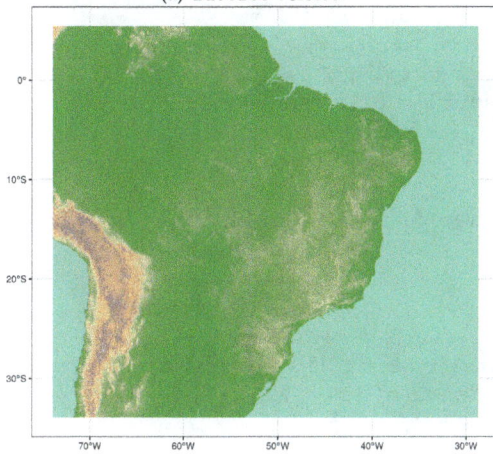

(b) `ggplot2` version

FIGURE 13.1: Digital elevation model of Brazil.

```
    sp.polygons(brazilAdmsp, col = 'black', lwd = 0.2)
    ## Centroids of administrative boundaries ...
    panel.points(centroids, col = 'black')
    ## ... with their labels
    panel.text(centroids, labels = admNames, pos = 3,
             cex = 0.7, fontfamily = 'Palatino', lineheight=.8)
    ## Country name
    panel.text(xyBrazil[1], xyBrazil[2], label = 'B R A Z I L',
             cex = 1.5, fontfamily = 'Palatino', fontface = 2)
    })


## ggplot2 version
ggplot() +
  geom_spatraster(data = brazilDEM,
                  show.legend = FALSE) +
  scale_fill_whitebox_c("high_relief",
                        na.value = "aquamarine") +
  geom_sf(data = brazilAdm,
          col = "black",
          linewidth = 0.15,
          fill = "transparent") +
  geom_sf(data = brazilRiv,
          col = "darkblue",
          linewidth = 0.1) +
  geom_point(data = centroids,
                  aes(x = longitude,
                      y = latitude)) +
  geom_text_repel(data = centroids,
                      aes(x = longitude,
                          y = latitude,
                          label = admNames),
                  family = "Palatino") +
  geom_text(aes(xyBrazil[1], xyBrazil[2],
             label = 'B R A Z I L'),
          size = 7,
          family = 'Palatino') +
  theme_bw()
```

(a) `lattice` version

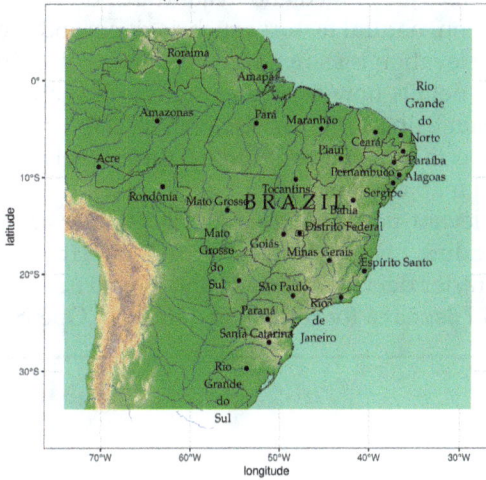

(b) `ggplot2` version

FIGURE 13.2: Physical map of Brazil. Main administrative regions are labeled.

Chapter 14

Reference Maps

A reference map focuses on the geographic location of features. In these maps, cities are named and major transport routes are identified. In addition, natural features such as rivers and mountains are named, and elevation is shown using a simple color shading.

This chapter details how to create a reference map of a northern region of Spain using data from OpenStreetMap.

The next subjects are covered in this chapter: label positioning, hill shading, and map overlaying.

The most relevant packages used in this chapter are: `raster` and `terra` for reading raster data; `rasterVis` and `tidyterra` for visualization of raster data; `colorspace` and `RColorBrewer` for palettes of colors; and `osmdata` for retrieving data from OpenStreetMap.

14.1 Introduction

Although I was born in Madrid, Galicia (north of Spain) is a very special region for me. More precisely, the Cedeira and Valdoviño regions offer a wonderful combination of wild sea, secluded beaches, and forests. I will show you a map of these marvelous places.

DOI: 10.1201/9781003485384-14

14.1.1 Packages

Several packages are employed to produce the map:

- `raster` and `terra` for reading and manipulating raster data.

- `sp` and `sf` for reading and manipulating spatial vector data.

```
library("raster")
library("terra")

library("sf")
library("sp")
```

- `rasterVis` and `tidyterra` for creating maps with `lattice` and `ggplot2`, respectively.

- `RColorBrewer` and `colorspace` for defining the palette of colors.

```
library("rasterVis")
library("tidyterra")

library("colorspace")
```

- `osmdata` for retrieving data from OpenStreetMap.

```
library("osmdata")
```

14.2 Retrieving Data from OpenStreetMap

The first step is to acquire information from the OpenStreetMap (OSM) project. There are several packages to extract data from this service, but while most of them only provide already rendered raster images, the osmdata package enables the use of the raw data with classes of the packages sp and sf.

The osmdata package obtains data from the overpass API[1], a read-only API that serves up custom selected parts of the OSM map data. The first step is specifying the bounding box with the function opq:

---

[1]http://www.overpass-api.de/

197

```
## Bounding box
xmin <- -8.1
ymin <- 43.62
xmax <- -8
ymax <- 43.7
cedeiraBB <- c(xmin = xmin, ymin = ymin,
               xmax = xmax, ymax = ymax)

## Overpass query
cedeiraOPQ <- opq(cedeiraBB)
```

Next, the query is completed, adding the required features with add_-
osm_feature. The three main arguments of this function are the overpass
query defined with opq, the feature key, and the value of this feature. Finally, the information contained in the query can be obtained as a Spatial* object with osmdata_sp, or as a sf object with osmdata_sf. The result is a list with three components, osm_points, osm_lines, and osm_-
polygons, containing the respective spatial object.

For example, the next code obtains the residential streets in the region
as a list of Spatial* objects.

```
streetsOSM <- add_osm_feature(cedeiraOPQ,
                              key = "highway",
                              value = "residential")

streetsSP <- osmdata_sp(streetsOSM)

print(streetsSP)
```

```
Object of class 'osmdata' with:
                 $bbox : 43.62,-8.1,43.7,-8
        $overpass_call : The call submitted to the overpass API
                $meta : metadata including timestamp and version numbers
          $osm_points : 'sp' SpatialPointsDataFrame with 1470 points
           $osm_lines : 'sp' SpatialLinesDataFrame with 226 lines
        $osm_polygons : 'sp' SpatialPolygonsDataFrame with 7 polygons
       $osm_multilines : 'sp' SpatialNADataFrame with 0 multilines
    $osm_multipolygons : 'sp' SpatialPolygonsDataFrame with 0 multipolygons
```

Because this procedure is to be repeated several times, I define a wrapper function that provides a SpatialLinesDataFrame object or a Spatial-
PointsDataFrame object depending on the value of its argument type:

```r
spFromOSM <- function(source, key, value, type = 'lines')
{
    osm <- add_osm_feature(source, key, value)
    spdata <- osmdata_sp(osm)
    switch(type,
           lines = spdata$osm_lines,
           points = spdata$osm_points)
}
```

and a function that provides a `sf` object:

```r
sfFromOSM <- function(bb, key, value, type = 'lines')
{
    source <- opq(bb)
    osm <- add_osm_feature(source, key, value)
    sfdata <- osmdata_sf(osm)
    switch(type,
           lines = st_crop(sfdata$osm_lines, bb),
           points = sfdata$osm_points)

}
```

The next code uses these functions to obtain the different types of roads and streets in the region as `SpatialLinesDataFrame` objects:

```r
streetsSP <- spFromOSM(cedeiraOPQ,
                       key = "highway", value = "residential")
primarySP <- spFromOSM(cedeiraOPQ,
                       key = "highway", value = "primary")
secondarySP <- spFromOSM(cedeiraOPQ,
                         key = "highway", value = "secondary")
tertiarySP <- spFromOSM(cedeiraOPQ,
                        key = "highway", value = "tertiary")
unclassifiedSP <- spFromOSM(cedeiraOPQ,
                            key = "highway", value = "unclassified")
footwaySP <- spFromOSM(cedeiraOPQ,
                       key = "highway", value = "footway")
stepsSP <- spFromOSM(cedeiraOPQ,
                     key = "highway", value = "steps")
```

and `sf` objects:

```r
streetsSF <- sfFromOSM(cedeiraBB,
                       key = "highway", value = "residential")
primarySF <- sfFromOSM(cedeiraBB,
                       key = "highway", value = "primary")
```

```
secondarySF <- sfFromOSM(cedeiraBB,
                     key = "highway", value = "secondary")
tertiarySF <- sfFromOSM(cedeiraBB,
                     key = "highway", value = "tertiary")
unclassifiedSF <- sfFromOSM(cedeiraBB,
                        key = "highway", value = "unclassified")
footwaySF <- sfFromOSM(cedeiraBB,
                     key = "highway", value = "footway")
stepsSF <- sfFromOSM(cedeiraBB,
                   key = "highway", value = "steps")
```

A similar procedure can be applied to construct a SpatialPointsData-Frame object with the collection of places with names:

```
citySP <- spFromOSM(cedeiraOPQ, key = "place",
                 value = "town", type = "points")

placesHsp <- spFromOSM(cedeiraOPQ, key = "place",
                    value = "hamlet", type = "points")
placesHsp <- subset(placesHsp, as.numeric(population) > 30)
```

The sfFromOSM function retrieves the same information as sf objects:

```
citySF <- sfFromOSM(cedeiraBB, key = "place",
                 value = "town", type = "points")

placesHsf <- sfFromOSM(cedeiraBB, key = "place",
                    value = "hamlet", type = "points")
placesHsf <- subset(placesHsf, as.numeric(population) > 30)
```

14.3   Hill Shading

The second step is to produce layers to display the topography. A suitable method is shaded relief or hill shading, previously exposed in section 11.1.1.

The hill shade layer is computed from the slope and aspect layers derived from a Digital Elevation Model. The DEM of Galicia is available at the Geonetwork service of the Xunta de Galicia[2]. I have extracted the data corresponding to the region of interest using crop, and the corresponding files are available at the data folder of the book repository.

---

[2]http://xeocatalogo.xunta.es/geonetwork/srv/gl/main.home

```
projCedeira <- projection(citySP)

demCedeira <- raster('data/Spatial/demCedeira')
projection(demCedeira) <- projCedeira

## Crop the DEM using the bounding box of the OSM data
OSMextent <- extent(extendrange(c(xmin, xmax)),
                    extendrange(c(ymin, ymax)))
demCedeira <- crop(demCedeira, OSMextent)

## Discard values below sea level
demCedeira[demCedeira <= 0] <- NA
```

The slope and aspect layers are computed with the terrain function, and the hill shade layer is derived with these layers for a fixed sun position.

```
slope <- terrain(demCedeira, 'slope')
aspect <- terrain(demCedeira, 'aspect')
hsCedeira <- hillShade(slope = slope, aspect = aspect,
                       angle = 20, direction = 30)
```

As it has been explained in sections 11.1 and 11.1.1, this same procedure can be implemented with the terra package. It will not be repeated here. For simplicity, we will use the conversion function:

```
demCedeiraT <- rast(demCedeira)
hsCedeiraT <- rast(hsCedeira)
```

14.4   Overlaying Layers of Information

And finally, the third step is to display the different layers of information in the correct order:

The rasterVis package creates the hill shade layer with the levelplot method for Raster objects. The GrTheme is modified to display the sea region with blue color. The result is shown in Figure 14.4.

```
## The background color of the panel is set to blue to represent
    the
## sea
hsTheme <- GrTheme(panel.background = list(col = "skyblue3"))

hsLattice <- levelplot(hsCedeira, maxpixels = ncell(hsCedeira),
                       par.settings = hsTheme,
```

```
                    margin = FALSE, colorkey = FALSE,
                    xlab = "", ylab = "")
```

The `tidyterra` package displays the hill shade layer with the `geom_-spatraster` function.

```
greyPal <- rev(brewer.pal(n = 9, name = "Greys"))

ggplot() +
  geom_spatraster(data = hsCedeiraT,
                  show.legend = FALSE) +
  scale_fill_gradientn(colours = greyPal,
                       na.value = "skyblue3") +
  theme_bw()
```

However, this layer cannot be combined with the next graphics layer (the DEM layer) because both define the color scale. This problem can be solved thanks to the `fill` argument of the `geom_spatraster` function: when we pass a vector of colors to this argument, the function will create a layer without mapping to the color scale of the graphic. Therefore, we have to build the scale of this layer. The result is shown in Figure 14.4.

```
## Build a vector of greys based on the Color Brewer pal
greyRamp <- colorRampPalette(greyPal)
greys <- greyRamp(255) ##255 colors, 0 to 254

## Classify the values of the raster in 255 classes
idx <- classify(hsCedeiraT, 255, ## 255 cuts
                include.lowest = TRUE)
idx <- as.vector(idx)
## Map these classes to the vector of colors
palGreys <- greys[idx + 1] ## 1:255 for indexing

hsGGplot <- geom_spatraster(data = hsCedeiraT,
                fill = palGreys,
                ## Plot every cell of the raster
                maxcell = Inf)
```

The DEM raster is printed with terrain colors and semitransparency over the hill shade layer. The `lattice` version is shown in Figure 14.4 and the `ggplot2` version in Figure 14.4.

```
## DEM with terrain colors and semitransparency
terrainPal <- terrain_hcl(n = 15)
```

(a) `lattice` version.

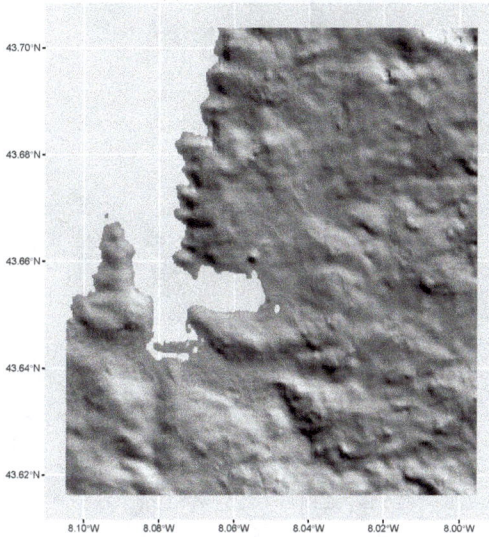

(b) `ggplot2` version.

FIGURE 14.1: Local topography of Cedeira (Galicia, Spain) displayed with the hill shading technique.

```
## Lattice version
terrainTheme <- rasterTheme(region = terrainPal,
                    regions = list(alpha = 0.6))

demLattice <- levelplot(demCedeira, maxpixels = ncell(demCedeira)
    ,
                    par.settings = terrainTheme,
                    margin = FALSE, colorkey = FALSE)

## ggplot version
demGGplot <- geom_spatraster(data = demCedeiraT,
                    alpha = 0.6,
                    show.legend = FALSE)
terrainScale <- scale_fill_gradientn(colours = terrainPal,
                    na.value = "skyblue3")
```

The roads are displayed with auxiliary functions (sp.road and sf_-
road) that produce a colored line over a thicker black line, and the towns
and villages and their names are displayed with the auxiliary functions
sp.places and sf_places.

```
## Auxiliary function to display the roads.

## A thicker black line in the background and a
## thinner one with an appropiate color.

## sp version
sp.road <- function(line, lwd = 6, blwd = 7,
                col = "indianred1", bcol = "black") {
  sp.lines(line, lwd = blwd, col = bcol)
  sp.lines(line, lwd = lwd, col = col)
}

## sf version
sf_road <- function(line, lwd = 1, blwd = 1.1,
                col = "indianred1", bcol = "black") {
  list(
    geom_sf(data = line, linewidth = blwd, col = bcol),
    geom_sf(data = line, linewidth = lwd, col = col)
  )
}

##Auxiliary function to display the towns and villages.
```

(a) `lattice` version.

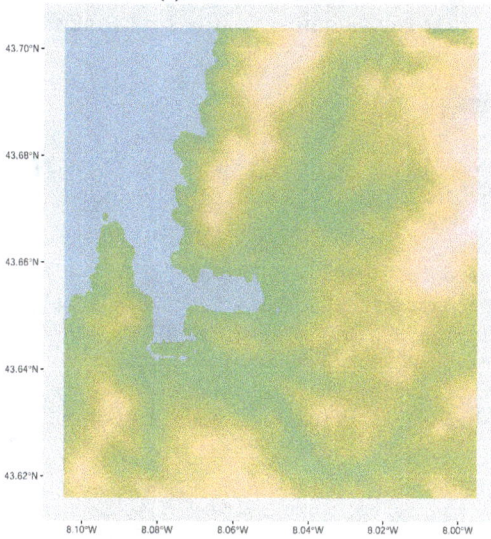

(b) `ggplot2` version.

FIGURE 14.2: Local topography of Cedeira (Galicia, Spain).

```r
## sp version
sp.places <- function(places, point.size= 0.4, text.size = 0.8) {
  sp.points(places, pch = 19, col = "black",
          cex = point.size, alpha = 0.8)
  sp.text(coordinates(places), places$name,
        pos = 3,
        fontfamily = "Palatino",
        cex = text.size, col = "black")
}


## sf version
sf_places <- function(places, text_size, point_size, vjust = -1)
{
  list(
    geom_sf(data = places, size = point_size),
    geom_sf_text(aes(label = name), data = places,
                size = text_size, vjust = vjust)
  )
}
```

The final figure is produced by superimposing the layers of sp objects with the +.trellis mechanism of the latticeExtra package (Figure 14.4):

```r
## Hill shade and DEM overlaid
hsLattice +
  demLattice +
  ## Roads and places
  layer({
    ## Street and roads
    sp.road(streetsSP, lwd = 1, blwd = 1, col = "white")
    sp.road(unclassifiedSP, lwd = 2, blwd = 2, col = "white")
    sp.road(footwaySP, lwd = 2, blwd = 2, col = "white")
    sp.road(stepsSP, lwd = 2, blwd = 2, col = "white")
    sp.road(tertiarySP, lwd = 4, blwd = 4, col = "palegreen")
    sp.road(secondarySP, lwd = 6, blwd = 6, col = "midnightblue")
    sp.road(primarySP, lwd = 7, blwd = 8, col = "indianred1")
    ## Places except Cedeira town
    sp.places(placesHsp, point.size = 0.4, text.size = 0.8)
    ## Cedeira town
    sp.places(citySP, point.size = 1.2, text.size = 1.5)
  })
```

or superimposing the layers of sf objects with the +.gg mechanism of ggplot2 (Figure 14.4):

```
ggplot() +
  hsGGplot +
  demGGplot + terrainScale +
  ## Street and roads
  sf_road(streetsSF, lwd = .4, blwd = .5, col = "white") +
  sf_road(unclassifiedSF, lwd = .4, blwd = .5, col = "white") +
  sf_road(footwaySF, lwd = .4, blwd = .5, col = "white") +
  sf_road(stepsSF, lwd = .4, blwd = .5, col = "white") +
  sf_road(tertiarySF, lwd = .8, blwd = .9, col = "palegreen") +
  sf_road(secondarySF, lwd = .9, blwd = 1, col = "midnightblue") +
  sf_road(primarySF, lwd = 1.1, blwd = 1.2, col = "indianred1") +
  ## Places
  sf_places(placesHsf, point_size = 1, text_size = 3) +
  sf_places(citySF, point_size = 3, text_size = 5) +
  theme_bw() + xlab("") + ylab("")
```

(a) `lattice` version.

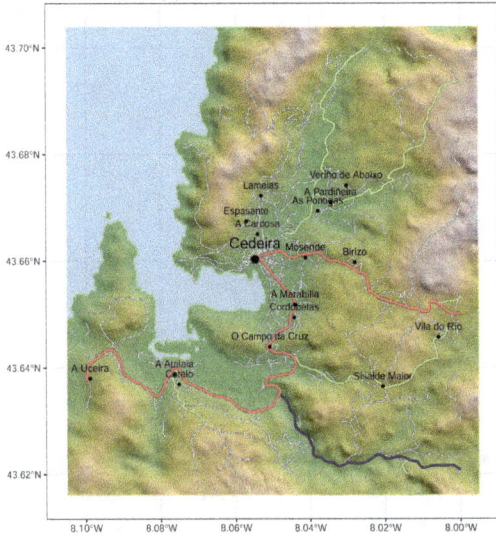

(b) `ggplot2` version.

FIGURE 14.3: Main roads near Cedeira, Galicia. Local topography is displayed with the hill shading technique. Some places are highlighted.

Chapter 15

About the Data

15.1 Air Quality in Madrid

Air pollution is harmful to health and contributes to respiratory and cardiac diseases and has a negative impact on natural ecosystems, agriculture, and the built environment. In Spain, the principal pollutants are particulate matter (PM), tropospheric ozone, nitrogen dioxide, and environmental noise[1].

The surveillance system of the Integrated Air Quality System of the Madrid City Council consists of twenty-four remote stations equipped with analyzers for gases (NO_X, CO, ozone, BT_X, HCs, SO_2) and particles (PM10, PM2.5), which measure pollution in different areas of the urban environment. In addition, many of the stations also include sensors to provide meteorological data.

15.1.1 Retrieve data

The detailed information of the network of measuring stations is available at the Open Data portal of Madrid[2]. The book repository contains a clean version of this file:

---

[1] https://www.eea.europa.eu/soer/2015/countries/spain

[2] The Open Data portal is https://datos.madrid.es/. The information of the measuring stations is available at https://datos.madrid.es/egob/catalogo/212629-1-estaciones-control-aire.csv.

```
airStations <- read.csv2("data/Spatial/airStations.csv")
head(airStations)
```

The air pollution data are available from the Madrid City Council web-page[3], and the structure of the file is documented in a PDF document[4]. The result after cleaning the data for the year 2016 is stored in the file `airQual-ity.csv`

```
airQuality <- read.csv2("data/Spatial/airQuality.csv")
head(airQuality)
```

15.1.2   Combine Data and Spatial Locations

Our next step is to combine the data and spatial information. The locations are contained in `airStations`, a `data.frame` that is converted to a `SpatialPointsDataFrame` object with the `coordinates` method.

```
library("sf")

## Spatial location of stations
airStations <- st_as_sf(airStations,
                        coords = c("long", "lat"),
                        crs = 4326)
```

On the other hand, the `airQuality data.frame` comprises the air quality daily measurements. We will retain only the NO_2 time series and represent each station with aggregated values (mean, median, and standard deviation) computed with `aggregate`:

```
NO2 <- subset(airQuality, codParam == 8)

NO2agg <- aggregate(dat ~ codEst, data = NO2,
                    FUN = function(x) {
                      c(mean = signif(mean(x), 3),
                        median = median(x),
                        sd = signif(sd(x), 3))
                    })
NO2agg <- do.call(cbind, NO2agg)
NO2agg <- as.data.frame(NO2agg)
```

---

[3]Use the search tool with the text "Calidad del aire. Datos diarios desde 2001".
[4]https://datos.madrid.es/FWProjects/egob/Catalogo/MedioAmbiente/Aire/
Ficheros/Interprete_ficheros_%20calidad_%20del_%20aire_global.pdf

The aggregated values (a `data.frame`) and the spatial information (a `sf` object) are combined with the `cbind` method of the `sf` package. Previously, the `data.frame` is reordered by matching against the shared key column (`airStations$Code` and `NO2agg$codEst`):

```
## Link aggregated data with stations to obtain a sf object
## Code and codEst are the stations codes
idxNO2 <- match(airStations$Code, NO2agg$codEst)
NO2sf <- cbind(airStations[, c("Name", "alt")],
               NO2agg[idxNO2, ])


## Save the result
st_write(NO2sf, dsn = "data/Spatial/", layer = "NO2sf",
         driver = "ESRI Shapefile")
```

15.2 Spanish General Elections

The results of the 2016 Spanish general elections[5] are available from the ministry webpage[6] and at the `data` folder of the book repository. Each region of the map will represent the percentage of votes (`pcMax`) obtained by the predominant political option (`whichMax`) at the corresponding municipality. Only six groups are considered: the four main parties (`PP`, `PSOE`, `UP`, and `Cs`), the abstention results (`ABS`), and the remaining parties (`OTH`). Each region will be identified by the `PROVMUN` code.

```
dat2016 <- read.csv("data/Spatial/GeneralSpanishElections2016.csv
    ")


population <- dat2016$Población
census <- dat2016$Total.censo.electoral
validVotes <- dat2016$Votos.válidos
## Election results per political party and municipality
votesData <- dat2016[, -(1:13)]
## Abstention as an additional party
votesData$ABS <- census - validVotes
## UP is a coalition of several parties
UPcols <- grep("PODEMOS|ECP", names(votesData))
votesData$UP <- rowSums(votesData[, UPcols])
votesData[, UPcols] <- NULL
```

---

[5]https://en.wikipedia.org/wiki/Spanish_general_election,_2016
[6]https://infoelectoral.interior.gob.es/es/elecciones-celebradas/area-de-descargas/

```r
## Winner party at each municipality
whichMax <- apply(votesData, 1, function(x)names(votesData)[
    which.max(x)])
## Results of the winner party at each municipality
Max <- apply(votesData, 1, max)
## OTH for everything but PP, PSOE, UP, Cs, and ABS
whichMax[!(whichMax %in% c("PP", "PSOE", "UP", "C.s", "ABS"))] <-
    "OTH"
## Percentage of votes with the electoral census
pcMax <- Max/census * 100

## Province-Municipality code. sprintf formats a number with
    leading zeros.
PROV <- sprintf("%02d", dat2016$Código.de.Provincia)
MUN <- sprintf("%03d", dat2016$Código.de.Municipio)
PROVMUN <- paste0(PROV, MUN)

votes2016 <- data.frame(PROV, MUN, PROVMUN,
                        population, census, validVotes,
                        whichMax, Max, pcMax)

write.csv(votes2016, "data/Spatial/votes2016.csv", row.names =
    FALSE)
```

15.2.1 Administrative Boundaries

The Spanish administrative boundaries are available as shapefiles at the National Statistics Institute (INE) of Spain webpage[7]. The municipalities, sfMun, are read with st_read of the sf package.

```r
library("sf")

old <- setwd(tempdir())

download.file("https://www.ine.es/pcaxis/mapas_completo_
    municipal.zip",
            "mapas_completo_municipal.zip")
unzip("mapas_completo_municipal.zip")

sfMun <- st_read("esp_muni_0109.shp", crs = 25830,
```

---

[7]http://www.ine.es/ > Products and services > Publications > Download the PC-Axis program > Municipal maps

```
                    stringsAsFactors = TRUE)
sfMun <- subset(sfMun, !is.na(sfMun$PROVMUN))

setwd(old)
```

The main step is to link the data with the polygons. The sf object behaves as a data.frame (with an additional column named geometry), and therefore the conventional methods for merging data.frames can be used here. The next code joins the sf object with the data.frame using the function match.

```
votes2016 <- read.csv("data/Spatial/votes2016.csv",
                 colClasses = c("factor", "factor", "factor",
                               "numeric", "numeric", "numeric",
                               "factor", "numeric", "numeric"))
## Match polygons and data with the PROVMUN column
idx <- match(sfMun$PROVMUN, votes2016$PROVMUN)

##Places without information
idxNA <- which(is.na(idx))

##Information to be added to the sf object
dat2add <- votes2016[idx, c("PROV", "population", "census", "
    validVotes",
                         "whichMax", "Max", "pcMax")]

## Spatial object with votes data
sfMapVotes <- cbind(sfMun, dat2add)

## Drop those places without information
sfMapVotes0 <- sfMapVotes[-idxNA, ]

## Save the result
st_write(sfMapVotes0, "data/Spatial/sfMapVotes0.shp")
```

Finally, Spanish maps are commonly displayed with the Canarian Islands next to the peninsula. First we have to extract the polygons of the islands and the polygons of the peninsula, and then shift the coordinates of the islands, summing the displacement on the geometry column. Finally, a new sf object binds the shifted islands with the peninsula.

```
## Extract Canarias islands from the sf object
canarias <- substr(sfMapVotes0$PROVMUN, 1, 2) %in% c("35", "38")
peninsula <- sfMapVotes0[!canarias,]
```

```
island <- sfMapVotes0[canarias,]

## Shift the island extent box to position them at the bottom
    right corner
dbbox <- st_bbox(peninsula) - st_bbox(island)
dxy <- dbbox[c("xmax", "ymin")]
island$geometry <- island$geometry + dxy

## Bind Peninsula (without islands) with shifted islands
st_crs(island) <- st_crs(peninsula)
sfMapVotes <- rbind(peninsula, island)

## Save the result
st_write(sfMapVotes, "data/Spatial/sfMapVotes.shp", append =
    FALSE)
```

15.3   Gross Domestic Product and Population

```
#
################################################################

## GDP and Population
#
################################################################
```

The INE also publishes data on population[8] and Gross Domestic Product (GDP)[9] for the provinces of Spain. The result after cleaning the data of population is stored in the file PopSpain.csv, and after filtering the data of GDP for the year 2020 is stored in the file GDPSpain2020.csv.

```
## Population of each province
popSpain <- read.csv("data/SpatioTime/PopSpain.csv")
popSpain2020 <- subset(popSpain, Year == 2020)
popSpain2020$PROV <- substring(popSpain2020$Province, 1, 2)

## GDP of each province
GDPSpain2020 <- read.csv("data/Spatial/GDPSpain2020.csv")
GDPSpain2020$PROV <- substring(GDPSpain2020$Province, 1, 2)
```

---

[8]https://www.ine.es/jaxiT3/Tabla.htm?t=2852&L=0
[9]https://www.ine.es/jaxi/Tabla.htm?tpx=67284&L=0

```
popGDPSpain2020 <- merge(popSpain2020, GDPSpain2020[, c("PROV", "
    GDP")])
```

The province boundaries are stored in the spain_provinces_2.shp file, retrieved from the webpage of the INE (Section 15.2.1).

```
library("sf")

sfProv <- st_read("data/Spatial/spain_provinces_2.shp", crs =
    25830,
              stringsAsFactors = TRUE)
```

Finally, the data frame and the provinces boundaries are merged together:

```
## Merge data with the polygons
sfPopGDPSpain <- merge(sfProv, popGDPSpain2020,
              by = "PROV")

st_write(sfPopGDPSpain, "data/Spatial/sfPopGDPSpain.shp")
```

15.4 CM SAF

The Satellite Application Facility on Climate Monitoring (CM SAF) is a joint venture of the Royal Netherlands Meteorological Institute, the Swedish Meteorological and Hydrological Institute, the Royal Meteorological Institute of Belgium, the Finnish Meteorological Institute, the Deutscher Wetterdienst, Meteoswiss, and the UK MetOffice, along with collaboration from the European Organization for the Exploitation of Meteorological Satellites (EUMETSAT) (CM SAF 2024). The CM-SAF was funded in 1992 to generate and store monthly and daily averages of meteorological data measured in a continuous way with a spatial resolution of 0.03° (15 kilometers). The CM SAF provides two categories of data: operational products and climate data. The operational products are built on data that are validated with on-ground stations and then provided in near-real-time to develop variability studies in diurnal and seasonal time scales. However, climate data are long-term data series to assess interannual variability (Posselt, Mueller, et al. 2012).

In this chapter we will display the annual average of the shortwave incoming solar radiation product (SIS) incident over Spain during 2008, computed from the monthly means of this variable. SIS collates shortwave radiation (0.2 to 4μm wavelength range) reaching a horizontal unit of Earth's surface obtained by processing information from geostationary satellites (METEOSAT) and also from polar satellites (MetOp and NOAA) (Schulz et al. 2009) and then validated with high-quality on-ground measurements from the Baseline Surface Radiation Network (BSRN)[10].

The monthly means of SIS are available upon request from the CM SAF webpage (Posselt, Müller, et al. 2011) and at the data folder of the book repository. Data from CM-SAF is published as raster files using the NetCDF format. The raster package provides the stack function to read a set of files and create a RasterStack object, where each layer stores the content of a file. Therefore, the twelve raster files of monthly averages produce a RasterStack with twelve layers.

```r
library("raster")

tmp <- tempdir()
unzip("data/Spatial/SISmm2008_CMSAF.zip", exdir = tmp)
filesCMSAF <- dir(tmp, pattern = "SISmm")
SISmm <- stack(paste(tmp, filesCMSAF, sep = "/"))
## CM-SAF data is average daily irradiance (W/m2). Multiply by 24
## hours to obtain daily irradiation (Wh/m2)
SISmm <- SISmm * 24
```

The RasterLayer object with annual averages is computed from the monthly means and stored as a NetCDF file.

```r
## Monthly irradiation: each month by the corresponding number of
      days
daysMonth <- c(31, 29, 31, 30, 31, 30, 31, 31, 30, 31, 30, 31)
SISm <- SISmm * daysMonth / 1000 ## kWh/m2
## Annual average
SISav <- sum(SISm)/sum(daysMonth)
writeRaster(SISav, file = "data/Spatial/SISav.nc")
```

15.5   Land Cover and Population Rasters

NASA's Earth Observing System (EOS)[11] is a coordinated series of polar-orbiting and low-inclination satellites for long-term global observations of the land surface, biosphere, solid Earth, atmosphere, and oceans. NEO-NASA[12], one of the projects included in EOS, provides a repository of

---

[10]https://bsrn.awi.de/
[11]http://eospso.gsfc.nasa.gov/
[12]https://neo.gsfc.nasa.gov/

global data imagery. We use the population density and land cover classification rasters. Both rasters must be downloaded from their respective
webpages as Geo-TIFF files.

```
library("raster")
## https://neo.gsfc.nasa.gov/view.php?datasetId=SEDAC_POP
pop <- raster("data/Spatial/875430rgb-167772161.0.FLOAT.TIFF")
## https://neo.gsfc.nasa.gov/view.php?datasetId=MCD12C1_T1
landClass <- raster("data/Spatial/241243rgb-167772161.0.TIFF")
```

Part III

Space-Time Data

Chapter 16

Displaying Spatiotemporal Data: Introduction

Space-time datasets are indexed in both space and time. The data may consist of a spatial vector object (for example, points or polygons) or raster data at different times. The first case is representative of data from fixed sensors providing measurements abundant in time but sparse in space. The second case is the typical format of satellite imagery, which produces high spatial resolution data sparse in time (E. Pebesma 2012).

There are several visualization approaches to space-time data trying to cope with the four dimensions of the data (Cressie and C. Wikle 2015).

On the one hand, the data can be conceived as a collection of snapshots at different times. These snapshots can be displayed as a sequence of frames to produce an animation or can be printed on one page with different panels for each snapshot using the small-multiple technique described repeatedly in previous chapters.

On the other hand, one of the two spatial dimensions can be collapsed through an appropriate statistic (for example, mean or standard deviation) to produce a space-time plot (also known as a Hovmöller diagram). The axes of this graphic are typically longitude or latitude as the x-axis and time as the y-axis, with the value of the spatial-averaged value of the raster data represented with color.

DOI: 10.1201/9781003485384-16

Finally, the space-time object can be reduced to a multivariate time series (where each location is a variable or column of the time series) and displayed with the time series visualization techniques described in Part I. This approach is directly applicable to space-time data sparse in space (for example, point measurements at different times). However, it is mandatory to use aggregation in the case of raster data. In this case, the multivariate time series is composed of the evolution of the raster data averaged along a certain direction.

The next chapters, focused on areal space-time data (Chapter 18), raster space-time data (Chapter 19), and point space-time data (Chapter 17), illustrate with examples how to produce multipanel graphics, Hovmöller diagrams, time-series, and animations with R. Last, there is Chapter 20 with additional animations that are not restricted to the evolution of a variable over time.

16.1 Packages

The CRAN Tasks View "Handling and Analyzing Spatiotemporal Data" [1] summarizes the packages for reading, visualizing, and analyzing space-time data. This section provides a brief introduction to the `spacetime`, `sftime`, `raster`, `terra`, `rasterVis`, `rgl`, and `gganimate` packages. Most of the information has been extracted from their vignettes, webpages, and help pages. You should read them for detailed information.

16.1.1 spacetime

The `spacetime` package (E. Pebesma 2012) is built upon the classes and methods for spatial data from the `sp` package, and for time series data from the `xts` package. It defines several classes to represent four space-time layouts. In the context of this book, the most relevant are STF and STFDF, full space-time grids of observations for spatial features and observation time, with all space-time combinations.

Moreover, `spacetime` provides several methods for the following classes:

- `stConstruct` and `STFDF` create objects from single or multiple tables.

- `as` coerces to other spatiotemporal objects, `xts`, `Spatial*`, `matrix`, or `data.frame`.

---

[1] https://cran.r-project.org/view=SpatioTemporal

- `[[` selects or replaces data values.

- `[` selects spatial or temporal subsets and data variables.

- `over` retrieves index or data values of one object at the locations and times of another.

- `aggregate` aggregates data values over particular spatial, temporal, or spatiotemporal domains.

- `stplot` creates spatiotemporal plots. It is able to produce multi-panel plots, space-time plots, animations, and time series plots.

16.1.2 sftime

`sftime` provides time extension for simple features in R as defined in the `sf` package. Thus, it allows to store spatial features that are accompanied by time information.

An `sftime` object is an `sf` object with an additional time column that contains the temporal information. The `st_time` function gets or sets this time column. All subsetting methods for `sf` objects (Section 7.1.2) also work for `sftime` objects. These objects can be displayed with `ggplot2` and the `geom_sf` function.

16.1.3 raster

The `raster` package (R. J. Hijmans 2024a) is able to add time information associated with layers of a `RasterStack` or `RasterBrick` object with the `setZ` function. This information can be extracted with `getZ`.

If a `Raster*` object includes this information, the `zApply` function can be used to apply a function over a time series of layers of the object.

16.1.4 terra

`terra` (R. J. Hijmans 2024b) provides the `time` method to get or set the time of the layers of a `SpatRaster`. Besides, `fillTime` can add empty layers in between existing layers to ensure that the time step between layers is constant, and `mergeTime` combines multiple rasters into a single time series. Finally, the `*app` family of methods applies a function across layers or to groups of layers.

16.1.5 rasterVis

rasterVis (O. Perpiñán and R. Hijmans 2023) provides three methods to display spatiotemporal rasters:

1. hovmoller produces Hovmöller diagrams (Hovmöller 1949). The axes of this kind of diagram are typically longitude or latitude (x-axis) and time (ordinate or y-axis), with the value of some aggregated field represented through color. However, the user can define the direction with dirXY and the summary function with FUN.

2. horizonplot creates horizon graphs (Few 2008), with many time series displayed in parallel by cutting the vertical range into segments and overplotting them with color representing the magnitude and direction of deviation. Each time series corresponds to a geographical zone defined with dirXY and averaged with zonal.

3. xyplot displays conventional time series plots. Each time series corresponds to a geographical zone defined with dirXY and aggregated with zonal.

On the other hand, the histogram, densityplot, and bwplot methods accept a FUN argument to be applied to the z slot of Raster* object (defined by setZ). The result of this function is used as the grouping variable of the plot to create different panels.

16.1.6 rgl

rgl is a package that produces real-time interactive 3D plots. It allows you to interactively rotate, zoom the graphics, and select regions. This package uses the OpenGL[2] library as the rendering backend, providing an interface to graphics hardware. It contains high-level graphics functions similar to base R graphics but working in three dimensions. Moreover, it provides low level functions inspired by the grid package.

16.1.7 gganimate

The gganimate package expands the ecosystem of ggplot2, focused on static graphics, with new grammar classes for animation. The most relevant for this chapter are: transition_time, which defines how the data

---

[2]https://www.opengl.org/

evolves over time, `ease_aes`, which defines how the aesthetics change with transitions; and `shadow_wake`, which defines how to render data from past instants.

The animation is initiated with a conventional call to `ggplot` and the corresponding `geom_*` and `scale_*` functions, and the `animate` function renders the animation to a file.

16.2   Further Reading

- (Cressie and C. Wikle 2015) is a systematic approach to key quantitative techniques in statistics for spatiotemporal data. The book begins with separate treatments of temporal data and spatial data and later combines these concepts to discuss spatiotemporal statistical methods. There is a chapter devoted to exploratory methods, including visualization techniques.

- (E. Pebesma 2012) presents the `spacetime` package, which implements a set of classes for spatiotemporal data. This paper includes examples that illustrate how to import, subset, coerce, and export spatiotemporal data; proposes several visualization methods; and discusses spatiotemporal geostatistical interpolation.

- (Slocum 2022) (previously cited in Chapter 7.2) includes a chapter about map animation, discussing several approaches for displaying spatiotemporal data.

- (Hengl 2009) (previously cited in Chapter 7.2) includes a working example with spatiotemporal data to illustrate space-time variograms and interpolation.

- (Harrower and Fabrikant 2008) explore the role of animation in geographic visualization and outline the challenges, both conceptual and technical, involved in the creation and use of animated maps.

- The CRAN Tasks View "Handling and Analyzing Spatiotemporal Data"[3] summarizes the packages for reading, visualizing, and analyzing space-time data. The R-SIG-Geo mailing list [4] is a powerful resource for obtaining help.

---

[3]`https://cran.r-project.org/view=SpatioTemporal`
[4]`https://stat.ethz.ch/mailman/listinfo/R-SIG-Geo/`

Chapter 17

Spatiotemporal Point Observations

Throughout this chapter we will revisit the data from the Integrated Air Quality System of the Madrid City Council (Section 15.1) to illustrate visualization methods applicable for point space-time data. This dataset comprises the time series of measurements acquired at each station of the network during 2011. In Section 8, the data were converted from spatiotemporal data to spatial data, where the time information was suppressed to display only the yearly average values. In this chapter we will work with the whole space-time dataset using the tools provided by the spacetime package (E. Pebesma 2012).

Additionally, the time evolution of the data is represented with an animation where the radius and color of dynamic bubbles vary over time.

Most relevant packages in this chapter are: sp, zoo, spacetime, and sftime for reading and defining point space-time data; reshape2 for converting from long to wide format; and gridSVG for creating an animation.

DOI: 10.1201/9781003485384-17

17.1 Introduction

The starting point is to retrieve the data and combine it with the spatial and temporal information. The data are contained in the `airQuality` `data.frame`, and the locations are in `airStations`, another `data.frame`.

```
## Spatial location of stations
airStations <- read.csv2("data/Spatial/airStations.csv")
## Measurements data
airQuality <- read.csv2("data/Spatial/airQuality.csv")
## Only interested in NO2
NO2 <- airQuality[airQuality$codParam == 8, ]
```

There are two different solutions:

- the sp and spacetime packages (with the aid of zoo and reshape2),

- the sf and sftime packages.

```
library("zoo")
library("reshape2")

library("sp")
library("spacetime")

library("sf")
library("sftime")
```

17.2 Data processing

17.2.1 sp **and** spacetime

The `airStations data.frame` is converted to a `SpatialPointsDataFrame` object with the `coordinates` method.

```
airStationsSP <- airStations
## rownames are used as the ID of the Spatial object
rownames(airStationsSP) <- substring(airStationsSP$Code, 7)
coordinates(airStationsSP) <- ~ long + lat
proj4string(airStationsSP) <- CRS("+proj=longlat +ellps=WGS84")
```

Each row of the `NO2 data.frame` corresponds to a measurement at one of the stations during a day of the year (long format, following the schema proposed in (E. Pebesma 2012)). The `spacetime` package defines several classes for spatiotemporal data inheriting the classes defined by the sp and xts packages. In particular, the STFDF, a class for spatiotemporal data with full space-time grids with n spatial locations and m times, requires a data.frame with n·m rows (spatial index moving faster than temporal index). Thus, we need to transform this structure to build a multivariate time series where each station is a different variable (space-wide under the schema of (E. Pebesma 2012)). The procedure is:

- Add a column with the POSIXct time index (line 1).

- Reshape the data.frame from long to wide format with dcast (line 4).

- Define a multivariate time series with zoo (Figure 17.3, line 8).

- Coerce this time series to a vector with n·m rows (line 10).

- Finally, create the STFDF object with the previous components (line 12).

```
1   NO2$time <- as.Date(with(NO2,
2                        ISOdate(year, month, day)))
3
4   NO2wide <- dcast(NO2[, c('codEst', 'dat', 'time')],
5               time ~ codEst,
6               value.var = "dat")
7
8   NO2zoo <- zoo(NO2wide[,-1], NO2wide$time)
9
10  dats <- data.frame(vals = as.vector(t(NO2zoo)))
11
12  NO2st <- STFDF(sp = airStationsSP,
13              time = index(NO2zoo),
14              data = dats)
```

17.2.2 sf and sftime

The procedure with the sf and sftime packages is simpler.

First, the airStations data.frame is converted to a sf object with st_as_sf:

```
airStationsSF <- st_as_sf(airStations,
                    coords = c("long", "lat"),
                    crs = 4326)
```

Then, the time information is added to the NO2 object and is combined with `airStationsSF` thanks to `st_sftime`.

```
NO2$time <- as.Date(with(NO2,
                    ISOdate(year, month, day)))

idx <- match(NO2$codEst, airStationsSF$Code)

NO2sft <- st_sftime(dat = NO2$dat, code = NO2$codEst,
              geometry = airStationsSF$geometry[idx],
              time = NO2$time)
```

17.3 Graphics

The `stplot` function of the `spacetime` package supplies the main visualization methods for spatiotemporal data. When the mode `xy` is chosen (default), it is mainly a wrapper around `spplot` and displays a panel with the spatial data for each element of the time index (Figure 17.1). The problem with this approach is that only a limited number of panels can be correctly displayed on one page. In this example, we print the first twelve days of the sequence.

```
airPal <- colorRampPalette(c("springgreen1", "sienna3", "gray5"))
       (5)

stplot(NO2st[, 1:12],
     cuts = 5,
     col.regions = airPal,
     main = "",
     edge.col = "black")
```

The ggplot2 version of this figure uses the `geom_sf` function and produces a panel for each day with `facet_wrap`.

```
airPal <- colorRampPalette(c("springgreen1", "sienna3", "gray5"))
       (5)

ggplot(NO2sft) +
  geom_sf(aes(color = dat)) +
  facet_wrap(~ cut(time, 12)) +
  scale_colour_stepsn(n.breaks = 6, colours = airPal)
```

With the mode `xt`, the `stplot` function plots a space-time plot with space on the x-axis and time on the y-axis (Figure 17.2).

229

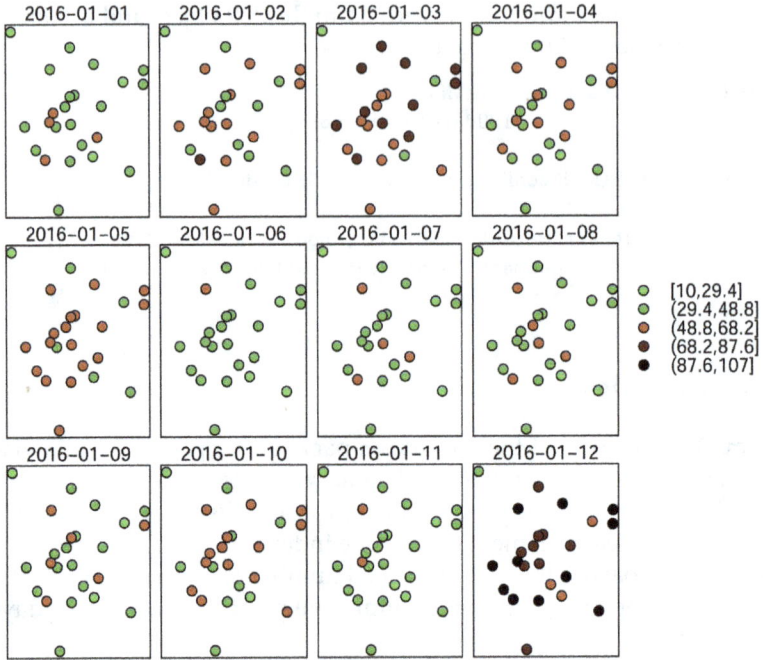

FIGURE 17.1: Scatterplots of the NO_2 values (2011) with a panel for each day of the time series. Each circle represents a different station.

```
stplot(NO2st, mode = "xt",
       col.regions = colorRampPalette(airPal)(15),
       scales = list(x = list(rot = 45)),
       ylab = "", xlab = "", main = "")
```

The ggplot2 version of this figure is produced with geom_tile, with the station codes on the x-axis and time on the y-axis:

```
ggplot(NO2sft) +
  geom_tile(aes(x = as.factor(code),
                y = time,
                fill = dat)) +
  scale_fill_stepsn(colours = airPal,
                    n.breaks = 6) +
  xlab("Station Code") + ylab("") +
  guides(x = guide_axis(angle = 45))
```

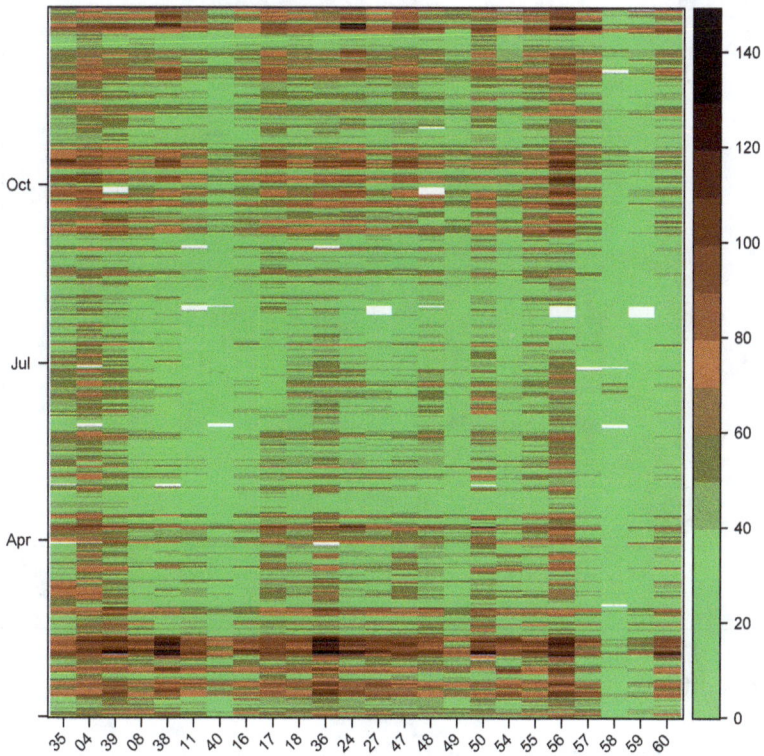

FIGURE 17.2: Space-time graphic of the NO_2 time series. Each column represents a different station (denoted with the last two digits of the code).

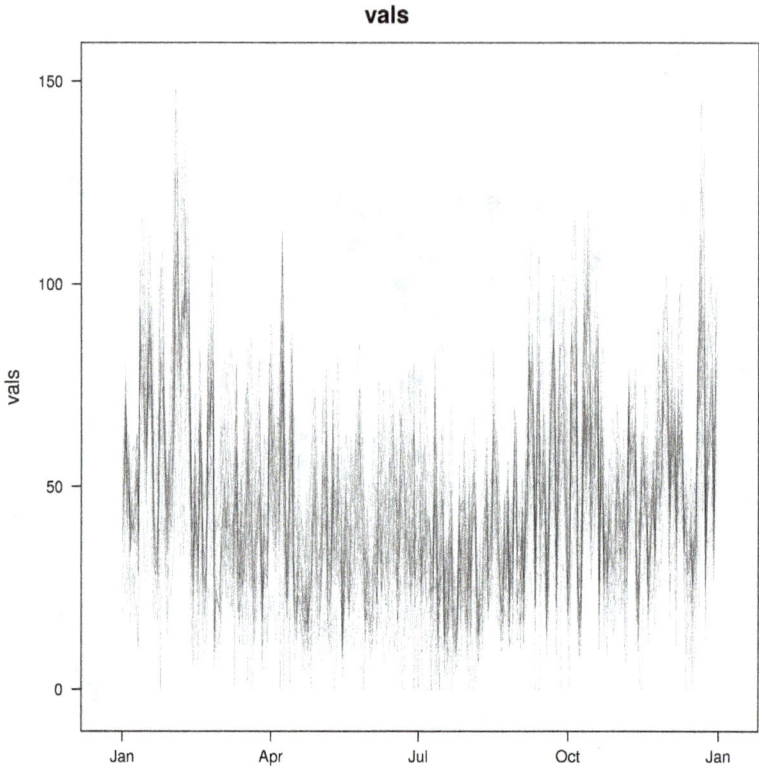

FIGURE 17.3: Time graph of the NO_2 time series (2011). Each line represents a different station.

Finally, if `stplot` is called with the mode `ts`, data are coerced to a multivariate time series that is displayed in a single plot (Figure 17.3).

```
stplot(NO2st, mode = "ts",
       xlab = "",
       lwd = 0.1, col = "black", alpha = 0.6,
       auto.key = FALSE)
```

This figure can be recreated with ggplot2 using `geom_line`, with time on the x-axis and measurements on the y-axis:

```
ggplot(NO2sft) +
  geom_line(aes(x=time, y = dat),
```

```
        colour = "black",
        linewidth = 0.25,
        alpha = 0.6) +
theme_bw() + xlab("")
```

These three graphics complement each other and together provide a more complete view of the behavior of the data. For example, in Figure 17.1, we can find stations whose levels remain almost constant throughout the twelve-days period (namely, El Pardo-28079058[1], the station at the top-left corner that is far from the city center), while others fluctuate notably during this same period (for example, Barajas-28079027 and Urb. Embajada-28079055, the two nearby stations at the right). On the other hand, Figure 17.2 loses the spatial information but gives a more comprehensive view of the evolution of the network of stations. The station El Pardo-28079058 is significantly below the rest of the stations during the whole year, with the station Pza. Fdez Ladreda-28079056 being the opposite. In between, the stations could be divided into two or three groups according to their levels. Regardless, the network of stations reaches maximum values during the first days of autumn and at the end of winter. These maxima are clearly displayed in Figure 17.3.

17.4 ♨Animation

This section details a method to create an animation with point data built over the functionalities of the gridSVG package. We will work with the NO2st object previously described in section 17.1.

17.4.1 Initial Snapshot

The first step is to define the initial parameters of the animation: starting values and duration.

```
library("grid")
library("gridSVG")

## Initial parameters
start <- NO2st[,1]
## values will be encoded as size of circles,
## so we need to scale them
startVals <- start$vals/5000
```

---

[1]Use Figure 8.5 as a reference for the positions and codes of the stations.

```
nStations <- nrow(airStationsSP)
days <- index(NO2zoo)
nDays <- length(days)
## Duration in seconds of the animation
duration <- nDays*.3
```

The first snapshot of the data is produced with spplot. We define an auxiliary function, panel.circlesplot, to display the data encoding values with circles of variable size and color. This function uses grid.circle from the grid package.

The subsequent frames of the animation will modify the colors and sizes of the circles according to the NO2st object.

```
## Auxiliary panel function to display circles
panel.circlesplot <- function(x, y, cex, col = "gray",
                    name = "stationsCircles", ...) {
  grid.circle(x, y, r = cex,
            gp = gpar(fill = col, alpha = 0.5),
            default.units = "native", name = name)
}

pStart <- spplot(start, panel = panel.circlesplot,
            cex = startVals,
            scales = list(draw = TRUE), auto.key = FALSE)
pStart
```

17.4.2   Intermediate States to Create the Animation

From this initial state, grid.animate creates a collection of animated graphical objects with the intermediate states defined by animUnit and animValue. As previously stated, the NO_2 values will be encoded with the radius of each circle, and the color of the circles will distinguish between weekdays and weekends. The use of rep=TRUE ensures that the animation will be repeated indefinitely.

```
## Color to distinguish between weekdays ('green') and weekend
## ('blue')
isWeekend <- function(x) {format(x, "%w") %in% c(0, 6)}
color <- ifelse(isWeekend(days), "blue", "green")
colorAnim <- animValue(rep(color, each = nStations),
                id = rep(seq_len(nStations), nDays))
```

```
## Intermediate sizes of the circles
vals <- NO2st$vals/5000
vals[is.na(vals)] <- 0
radius <- animUnit(unit(vals, "native"),
                   id = rep(seq_len(nStations), nDays))

## Animation of circles including sizes and colors
grid.animate("stationsCircles",
             duration = duration,
             r = radius,
             fill = colorAnim,
             rep = TRUE)
```

17.4.3   Time Reference: Progress Bar

Information from an animation is better understood if a time reference
is included, for example, with a progress bar. The following code builds
a progress bar with ticks at the first day of each month and with color
changing from gray (background) to blue as the time advances. On the
other hand, it is convenient to provide a method so the user can stop and
restart the animation sequence if desired. This functionality is added with
the definition of two events, onmouseover and onmouseout, included with
the grid.garnish function.

```
## Progress bar
prettyDays <- pretty(days, 12)
## Width of the progress bar
pbWidth <- .95
## Background
grid.rect(.5, 0.01, width = pbWidth, height = .01,
          just = c("center", "bottom"),
          name = "bgbar", gp = gpar(fill = "gray"))

## Width of the progress bar for each day
dayWidth <- pbWidth/nDays
ticks <- c(0, cumsum(as.numeric(diff(prettyDays)))*dayWidth) +
    .025
grid.segments(ticks, .01, ticks, .02)
grid.text(format(prettyDays, "%d-%b"),
          ticks, .03, gp = gpar(cex = .5))
## Initial display of the progress bar
grid.rect(.025, .01, width = 0,
          height = .01, just = c("left", "bottom"),
```

```
            name = "pbar", gp = gpar(fill = "blue", alpha = ".3"))
## ...and its animation
grid.animate("pbar", duration = duration,
             width = seq(0, pbWidth, length = duration),
             rep = TRUE)
## Pause animations when mouse is over the progress bar
grid.garnish("bgbar",
             onmouseover = "document.documentElement.pauseAnimations
                 ()",
             onmouseout = "
                 document.documentElement.unpauseAnimations()")
```

The SVG file is finally produced with `grid.export` (Figure 17.4)

```
grid.export("figs/SpatioTime/NO2pb.svg")
```

17.4.4   Time Reference: A Time Series Plot

A different and more informative solution is to add a time series plot instead of a progress bar. This time series plot displays the average value of the set of stations, with a point and a vertical line to highlight the time position as the animation advances (Figure 17.5).

```
library("lattice")
library("latticeExtra")

## Time series with average value of the set of stations
NO2mean <- zoo(rowMeans(NO2zoo, na.rm = TRUE), index(NO2zoo))
## Time series plot with position highlighted
pTimeSeries <- xyplot(NO2mean, xlab = "", identifier = "timePlot"
    ) +
  layer({
    grid.points(0, .5, size = unit(.5, "char"),
                default.units = "npc",
                gp = gpar(fill = "gray"),
                name = "locator")
    grid.segments(0, 0, 0, 1, name = "vLine")
  })

print(pStart, position = c(0, .2, 1, 1), more = TRUE)
print(pTimeSeries, position = c(.1, 0, .9, .25))
```

Once again, `grid.animate` creates a sequence of intermediate states for each object of the graphical scenes: The signaling point and vertical

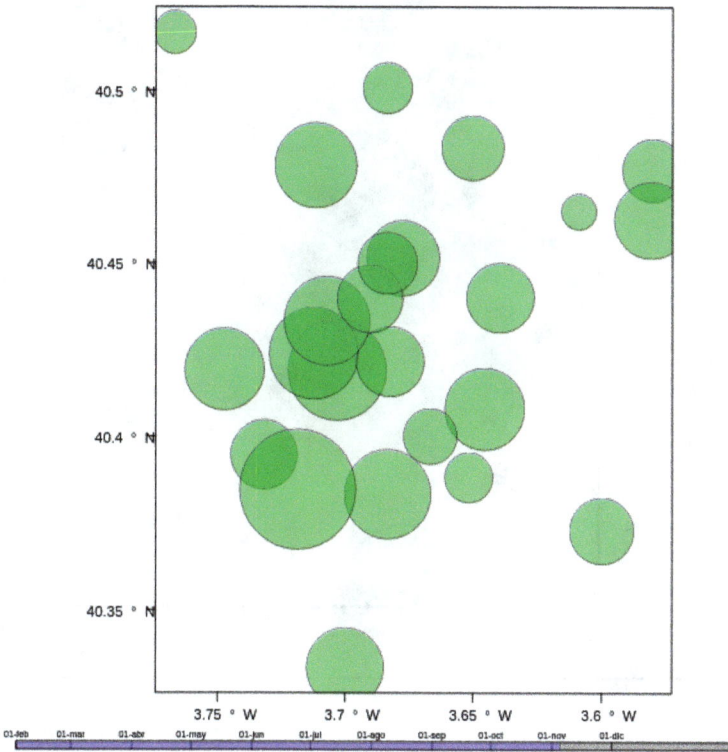

FIGURE 17.4: Animated circles of the NO_2 space-time data with a progress bar.

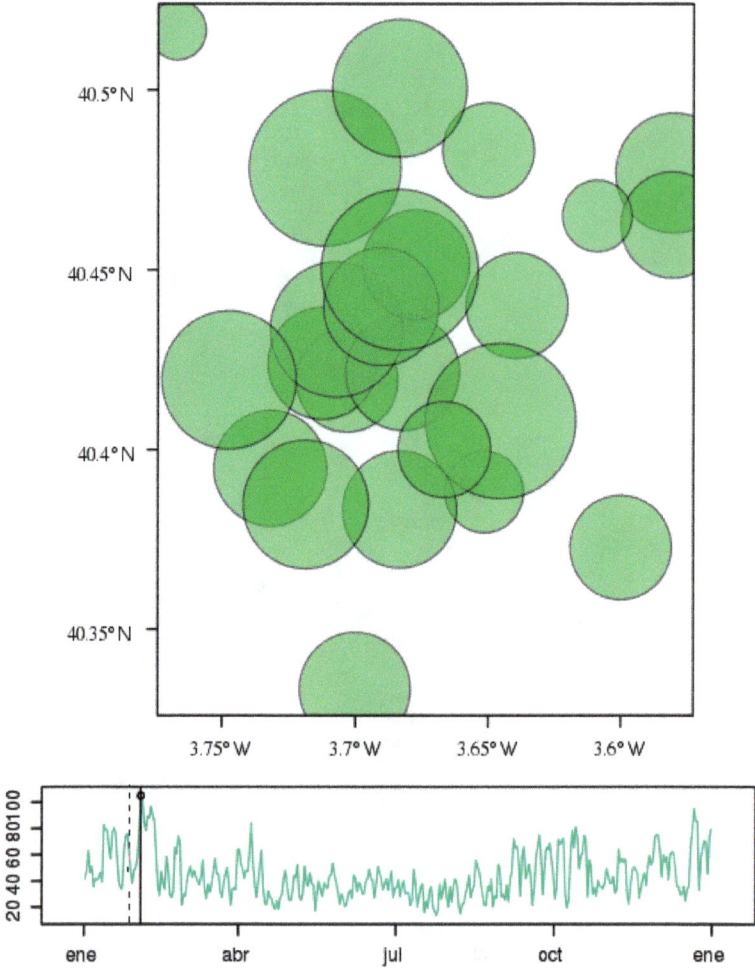

FIGURE 17.5: Animated circles of the NO_2 space-time data with a time series as reference.

line follow the time evolution, while the sizes and colors of each station circle change as in the previous approach. Moreover, the onmouseover and onmouseout events are defined with grid.garnish so the user can pause and restart the animation by hovering the mouse over the time series plot.

```r
grid.animate("locator",
             x = unit(as.numeric(index(NO2zoo)), "native"),
             y = unit(as.numeric(NO2mean), "native"),
             duration = duration, rep = TRUE)

xLine <- unit(index(NO2zoo), "native")

grid.animate("vLine",
             x0 = xLine, x1 = xLine,
             duration = duration, rep = TRUE)

grid.animate("stationsCircles",
             duration = duration,
             r = radius,
             fill = colorAnim,
             rep = TRUE)

## Pause animations when mouse is over the time series plot
grid.garnish("timePlot", grep = TRUE,
             onmouseover = "document.documentElement.pauseAnimations
                 ()",
             onmouseout = "
                 document.documentElement.unpauseAnimations()")

grid.export("figs/SpatioTime/vLine.svg")
```

Chapter 18

Spatiotemporal Areal Data

Choropleth and cartogram maps were covered in Chapters 9 and 10 without considering the evolution of the variables with time. This chapter employs these thematic maps with space-time variables both with static graphics and with animations.

Most relevant packages in this chapter are: sf for reading spatial data; ggplot2 for producing the static graphics; cartogram for computing the cartograms; and gganimate for creating the animations.

DOI: 10.1201/9781003485384-18

18.1 Introduction

The graphics presented in this chapter illustrate the progression of COVID-19 infections within the provinces of Spain during the period comprised between December 2021 and March 2022.

Choropleths and cartograms, two visualization techniques used in Chapters 9 and 10, are employed here. This chapter features animated versions of these techniques, in addition to static graphics based on the Trellis scheme (a rectangular array of plots).

This chapter relies on three key packages: sf, cartogram, and gganimate.

```r
library("sf")

library("cartogram")

library("gganimate")
```

18.2 Data

Data on COVID-19 infections have been adapted from the tables available on the website of the Institute of Health Carlos III[1] (Chapter 15.3):

```r
covidSpain <- read.csv("data/SpatioTime/covid.csv",
                    na.strings = NULL)
covidSpain$PROV <- sprintf("%02d", covidSpain$PROV)
covidSpain$day <- as.Date(covidSpain$day)

covidSpain <- subset(covidSpain,
                day < as.Date("2022-03-01") &
                day >= as.Date("2021-12-01"))
```

Infection numbers are usually standardized per 100,000 population. Population data are obtained from the National Institute of Statistics[2].

```r
## Population of each province
popSpain <- read.csv("data/SpatioTime/PopSpain.csv")
popSpain$PROV <- substring(popSpain$Province, 1, 2)
```

This analysis uses 2021 population data for its three-month duration.

---

[1] https://cnecovid.isciii.es/covid19/#documentaci%C3%B3n-y-datos
[2] https://www.ine.es/jaxiT3/Tabla.htm?t=2852

```
popSpain2021 <- subset(popSpain,
                       Year == 2021,
                       select = c(PROV, Population))

covidSpain <- merge(covidSpain, popSpain2021, by = "PROV")

## Number of cases per 100.000 population
covidSpain$cases_rel <- with(covidSpain,
                             num_cases/Population * 1e5)
```

Figure 18.1 is a level plot that depicts the progression of the COVID-19 infections in Spain[3] from December 2021 to March 2022.

```
ggplot(covidSpain) +
  geom_raster(aes(x = day,
                  y = PROV,
                  fill = cases_rel)) +
  scale_fill_distiller(palette = "PuBu", direction = 1) +
  theme_bw()
```

Finally, the covidSpain data frame and the provinces boundaries are merged together:

```
sfProv <- st_read("data/Spatial/spain_provinces_2.shp",
                  crs = 25830,
                  stringsAsFactors = TRUE)

## Merge data with the polygons
sfCovid <- merge(sfProv, covidSpain,
                 by = "PROV")
```

18.3 Choropleth

Choropleths with quantitative variables were extensively described in Section 9.2, but not in a time-dependent context. The representation of a space-time variable requires different approaches.

The static solution is the small multiples technique (section 4.1): a matrix of level plots where each cell depicts the COVID-19 infection rate in each province on a certain day. These multiple images displayed together allow the reader to compare the inter-frame differences and perceive the time evolution (Figure 18.2).

---

[3]The relation between the codes of the y-axis and the names of the provinces is available at this webpage: https://www.ine.es/daco/daco42/codmun/cod_provincia.htm

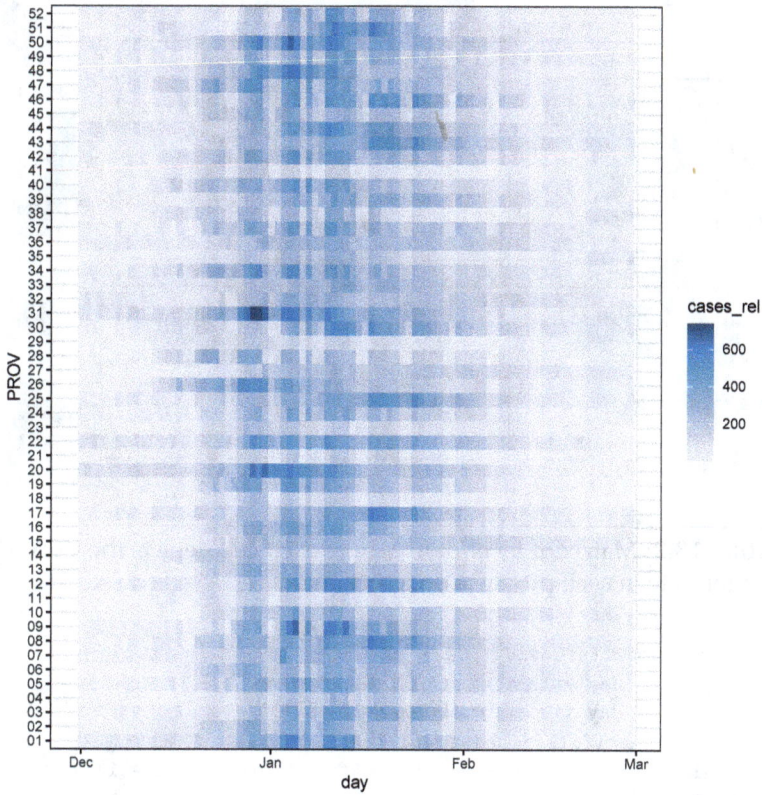

FIGURE 18.1: Level plot of the COVID-19 infections in Spain from December 2021 to March 2022. The y-axis represents the provinces by their codes.

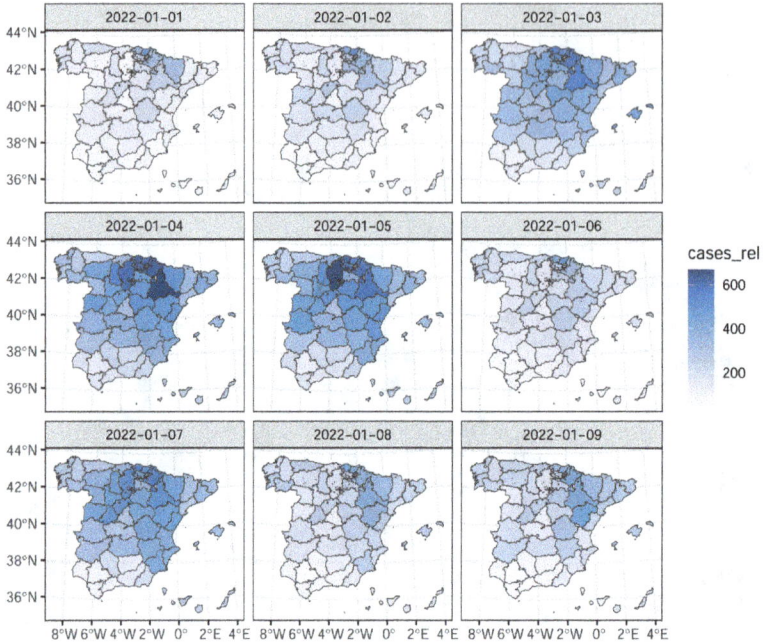

FIGURE 18.2: Matrix of level plots where each cell depicts the COVID-19 infection rate in each province on a certain day

```
ggplot(subset(sfCovid,
              day >= as.Date("2022-01-01") &
              day <= as.Date("2022-01-09"))) +
  geom_sf(aes(fill = cases_rel)) +
  scale_fill_distiller(palette = "PuBu", direction = 1) +
  theme_bw() +
  facet_wrap(~ day, nrow = 3)
```

The animated solution displays sequentially each of these level plots as frames of an animation. The gganimate package eases this task with transition_time, which can be added as a component of the ggplot call, and the animate function, which renders the animation. Figure 18.3 shows a snapshot of the result, available at the book website.

```
ggCovid <- ggplot(sfCovid) +
  geom_sf(aes(fill = cases_rel)) +
  scale_fill_distiller(palette = "PuBu", direction = 1) +
```

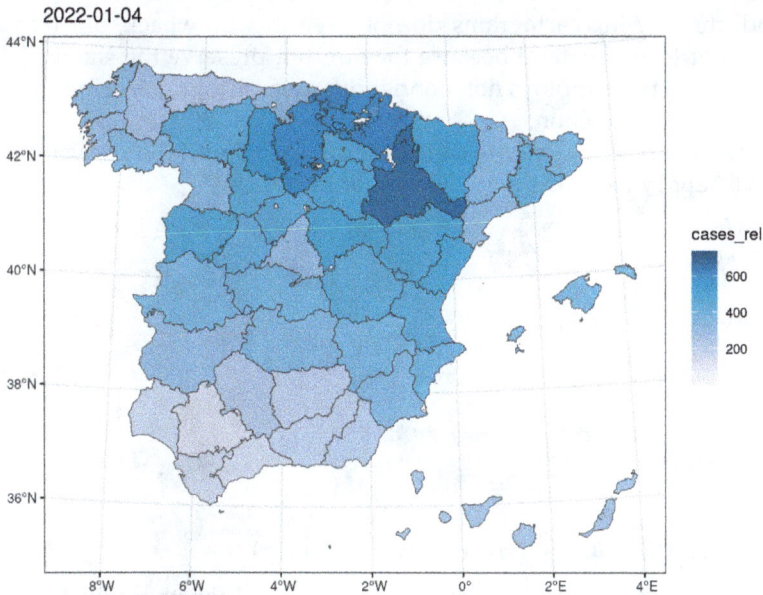

FIGURE 18.3: Snapshot of the animation of choropleths produced with gganimate and sf.

```
theme_bw() +
ggtitle("{format(frame_time, format = '%Y-%m-%d')}") +
transition_time(time = day)


animate(ggCovid,
        height = 1080, width = 1080,
        res = 150,
        units = "px")
```

18.4 Cartogram

Chapter 10 details how to obtain cartograms with the cartogram package. Three different types are available: non-contiguous area, contiguous area, and non-overlapping circles (Dorling). Contiguous area cartograms are discarded in this context because they are computationally expensive for space-time data due to the time required for transformation. On the other hand, the Dorling cartograms do not have this drawback, but they can be harder to understand because they do not preserve the shape. Therefore, this section employs non-contiguous area cartograms using the cartogram_ncont function.

The next code generates a cartogram for each day of the analysis period with a lapply loop. The result is a list of sf objects.

```r
fday <- as.Date("2022-01-01")
lday <- as.Date("2022-02-28")

days <- seq(fday, lday, by = "day")

cartCOVIDList <- lapply(days, function(d)
{
  x <- subset(sfCovid, day == d)
  cartogram_ncont(x,
                  weight = "cases_rel")
})
cartCOVID <- do.call(rbind, cartCOVIDList)
```

The collection of cartograms can be displayed together with the small multiples approach. The changes in the variable are represented both with the hue and with the size of the polygons (Figure 18.4).

```r
ggplot(subset(cartCOVID,
              day >= as.Date("2022-01-01") &
              day <= as.Date("2022-01-09"))) +
  geom_sf(data = sfProv) +
  geom_sf(aes(fill = cases_rel)) +
  scale_fill_distiller(palette = "PuBu", direction = 1) +
  theme_bw() +
  facet_wrap(~ day, nrow = 3)
```

These cartograms can also be displayed as frames of an animation with a ggplot call completed with transition_time and rendered with animate. Figure 18.5 shows a snapshot of the animation, available at the book website.

```r
ggCartCOVID <- ggplot(cartCOVID) +
  geom_sf(data = sfProv) +
  geom_sf(aes(fill = cases_rel, group = day)) +
  scale_fill_distiller(palette = "PuBu", direction = 1) +
  ggtitle("{format(frame_time)}") +
  transition_time(day) +
  theme_bw()
```

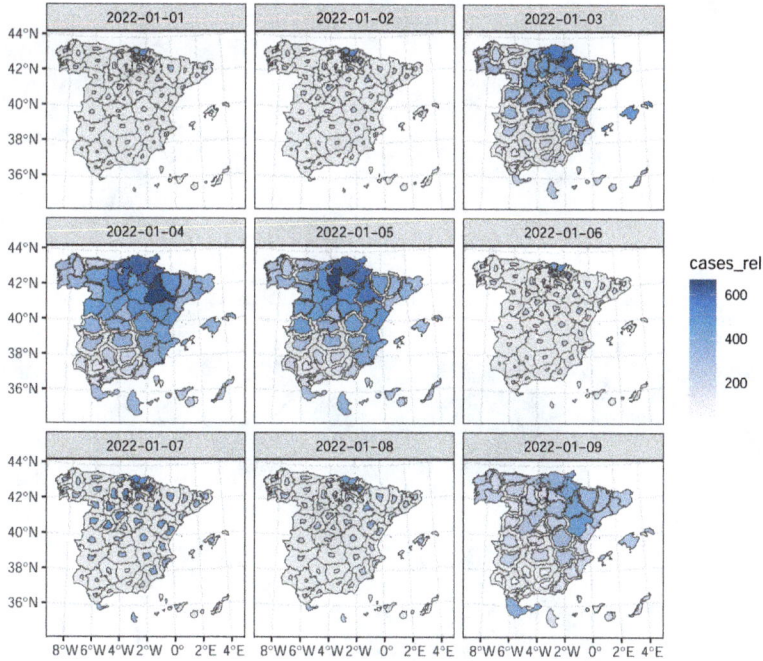

FIGURE 18.4: Matrix of non-contiguous area cartograms where each cell depicts the COVID-19 infection rate in each province on a certain day.

```
animate(ggCartCOVID,
        height = 1080, width = 1080,
        res = 150,
        units = "px")
```

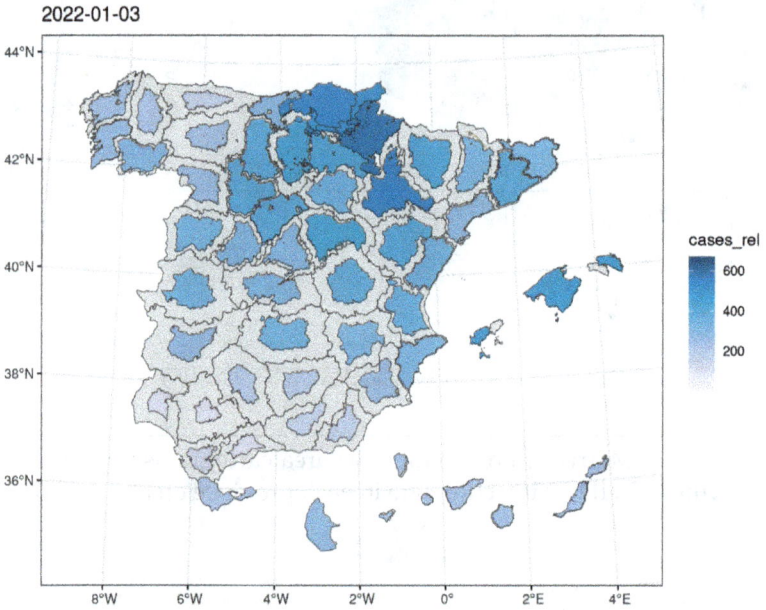

FIGURE 18.5: Snapshot of the animation of cartograms produced with gganimate and cartogram.

Chapter 19

Spatiotemporal Raster Data

A space-time raster dataset is a collection of raster layers indexed by time, or in other words, a time series of raster maps. The `raster` package defines the classes `RasterStack` and `RasterBrick` to build multilayer rasters. The index of the collection can be set with the function `setZ` (which is not restricted to time indexes). The `terra` provides the `time` function that gets or sets time information to the `SpatRaster` objects. The `rasterVis` package contains several methods to display space-time rasters with both packages.

Additionally, this chapter displays the individual layers of the space-time raster sequentially as movie frames to produce an animation.

This chapter covers these subjects: small multiples; scatterplot matrices; horizon graph; interactive and 3D visualization; and animations.

The most relevant packages in this chapter are: `raster` and `terra` for reading raster data; `zoo` for the definition of time series; `rasterVis` for displaying raster data; `mapview` for interactive graphics; and `rgl` for a 3D animation.

DOI: 10.1201/9781003485384-19

19.1 Introduction

Throughout this chapter we will work with a multilayer raster of daily solar radiation estimates from CM SAF (section 15.4) falling in the region of Galicia (north of Spain) during 2011. These data are arranged in a `Raster-Brick` with 365 layers using `brick` and time indexed with `setZ`.

```
library("raster")
library("terra")

library("zoo")

library("RColorBrewer")
library("rasterVis")

SISdm <- brick("data/SpatioTime/SISgal")

timeIndex <- seq(as.Date("2011-01-01"), by = "day", length = 365)
SISdm <- setZ(SISdm, timeIndex)
names(SISdm) <- format(timeIndex, "%a_%Y%m%d")
```

```
class        : RasterBrick
dimensions   : 70, 95, 6650, 365  (nrow, ncol, ncell, nlayers)
resolution   : 0.03, 0.03  (x, y)
extent       : -9.385, -6.535, 41.735, 43.835  (xmin, xmax, ymin, ymax)
coord. ref.  : +proj=longlat +datum=WGS84 +ellps=WGS84 +towgs84=0,0,0
data source  : data/SISgal.grd
names        : sáb_20110101, dom_20110102, lun_20110103, mar_20110104,
               mié_20110105, jue_20110106, vie_20110107, sáb_20110108,
               dom_20110109, lun_20110110, mar_20110111, mié_20110112,
               jue_20110113, vie_20110114, sáb_20110115,         ...
min values   :   22.368380,    17.167900,    14.132746,     6.816897,
                 11.616024,     4.851091,     7.251504,     7.717323,
                 32.492584,    16.786699,    20.433887,    25.984509,
                 24.190809,    25.485737,    31.256468,         ...
max values   :   88.84628,     93.43065,     84.44052,     81.79698,
                 47.08540,     49.01350,     80.98129,     74.12977,
                102.71733,     48.54120,     75.46757,     98.09704,
                104.65322,    104.29843,    107.23942,         ...
time         : 2011-01-01, 2011-12-31 (min, max)
```

The `terra` package is also able to include the timestamp in the objects with the `time` function:

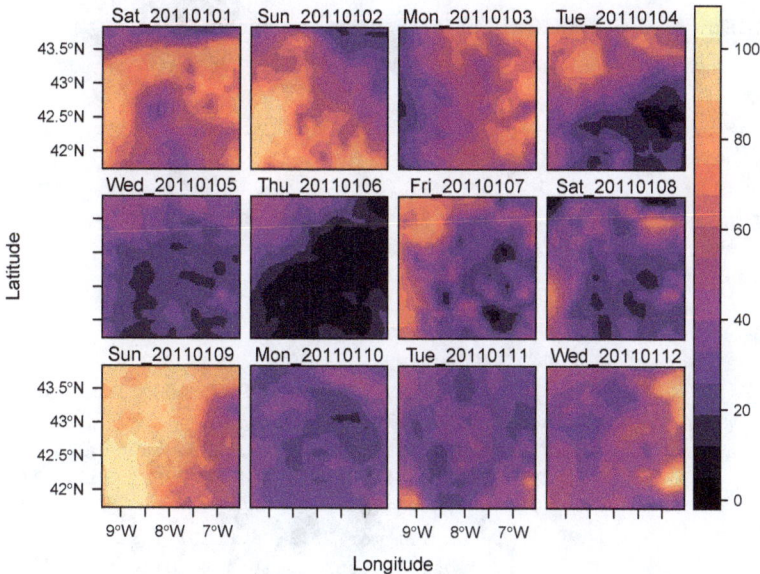

FIGURE 19.1: Level plot of daily averages of solar radiation.

```
SISdmt <- rast(SISdm)
time(SISdmt) <- timeIndex
```

The code of this chapter works with both the `RasterBrick` and the `SpatRaster` objects. For simplicity, only the `RasterBrick` object is used from here on.

19.2   Level Plots

This multilayer raster can be displayed with each snapshot in a panel using the small-multiple technique. The problem with this approach is that only a limited number of panels can be correctly displayed on one page. In this example, we print the first 12 days of the sequence (Figure 19.1).

```
levelplot(SISdm, layers = 1:12, panel = panel.levelplot.raster)
```

When the number of layers is very high, a partial solution is to aggregate the data, grouping the layers according to a time condition. For example, we can build a new space-time raster with the monthly averages using `zApply` and `as.yearmon`.

```
SISmm <- zApply(SISdm, by = as.yearmon, fun = 'mean')
```

```
class         : RasterBrick
dimensions    : 70, 95, 6650, 12  (nrow, ncol, ncell, nlayers)
resolution    : 0.03, 0.03  (x, y)
extent        : -9.385, -6.535, 41.735, 43.835  (xmin, xmax, ymin, ymax)
coord. ref.   : +proj=longlat +datum=WGS84 +ellps=WGS84 +towgs84=0,0,0
data source   : in memory
names         :  ene.2011,   feb.2011,   mar.2011,   abr.2011,   may.2011,
                 jun.2011,   jul.2011,   ago.2011,   sep.2011,   oct.2011,
                 nov.2011,   dic.2011
min values    :  45.79540,   90.73799, 127.99301, 206.37934, 171.51269,
                233.96372,  196.51799, 188.50950, 161.11302, 126.57236,
                 64.57646,   49.24799
max values    :  74.62326,  123.82005, 167.88226, 249.85596, 313.51773,
                340.51255,  334.61417, 273.41496, 225.10363, 164.54701,
                 82.47245,   79.40128
              : ene 2011, feb 2011, mar 2011, abr 2011, may 2011, jun 2011,
                jul 2011, ago 2011, sep 2011, oct 2011, nov 2011, dic 2011
```

This raster can be completely displayed on one page (Figure 19.2), although part of the information of the original data is lost in the aggregation procedure.

```
levelplot(SISmm, panel = panel.levelplot.raster)
```

19.3   Graphical Exploratory Data Analysis

There are other graphical tools that complement the previous maps. The scatterplot and the matrix of scatterplots, the histogram and kernel density plot, and the boxplot are among the most important tools in the frame of the Exploratory Data Analysis approach. Some of them were previously used with a spatial raster (Chapter 11). In this section we will use the histogram (Figure 19.3),

```
histogram(SISdm, FUN = as.yearmon)
```

the violin plot (a combination of a boxplot and a kernel density plot) (Figure 19.4),

```
bwplot(SISdm, FUN = as.yearmon)
```

and the matrix of scatterplots (section 4.1, Figure 19.5).

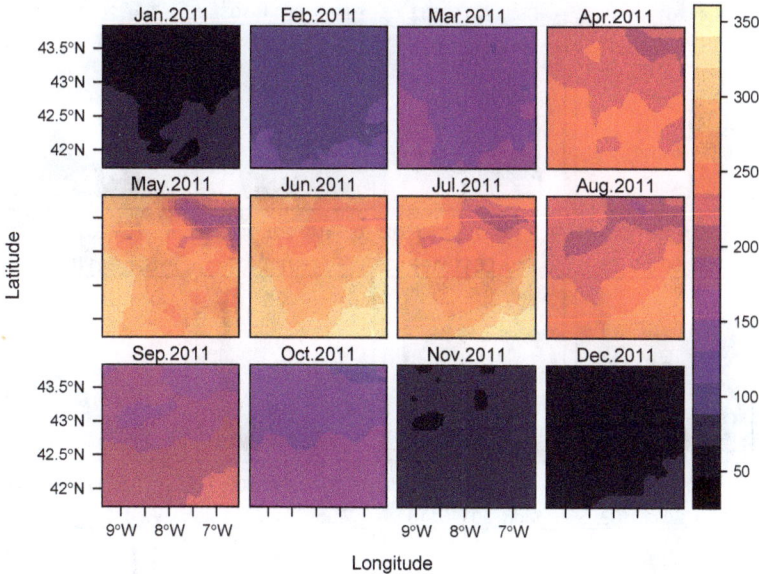

FIGURE 19.2: Level plot of monthly averages of solar radiation.

```
splom(SISmm, xlab = '', plot.loess = TRUE)
```

Both the histogram and the violin plot show that daily solar irradiation is bimodal almost every month. This is related to the predominance of clear sky and overcast days, with several partly cloudy days between these modes. This geographical region receives higher irradiation levels from June to September, and both the levels and the shape of the probability distribution contrast sharply with the winter.

The matrix of scatterplots displays a quasilinear relationship between the summer months due to the predominance of clear sky conditions. However, the relationships involving winter months become strongly nonlinear due to the presence of clouds.

19.4 Space-Time and Time Series Plots

The level plots of Figures 19.1 and 19.2 display the full 3D space-time data using a grid of panels where each layer is depicted in a separate panel. In section 19.5, this collection of layers will be displayed sequentially like

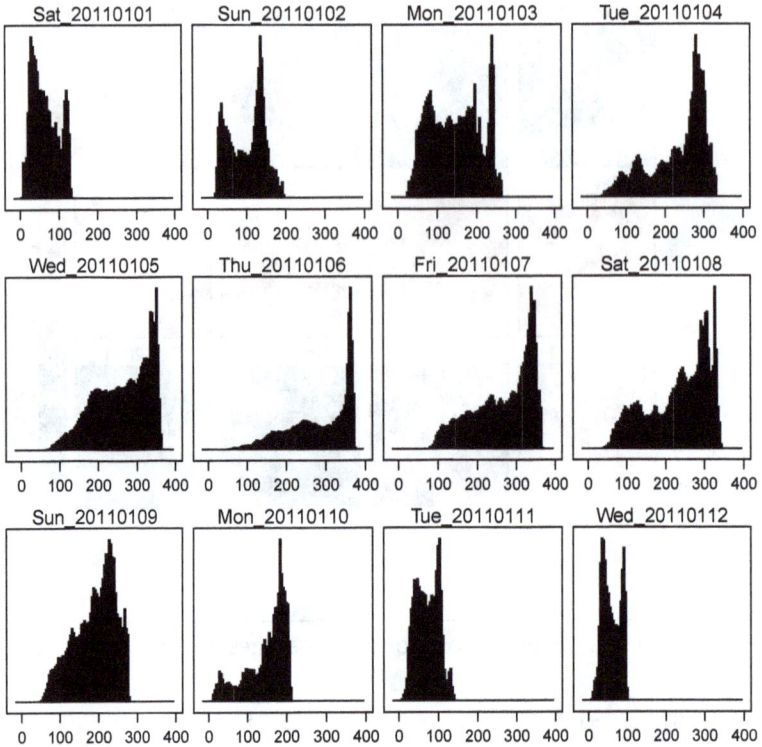

FIGURE 19.3: Histogram of monthly distribution of solar radiation.

frames of a movie to build an animation. In this section, the 3D raster is reduced to a 2D matrix with spatial aggregation following a certain direction. For example, Figure 19.6 displays with colors the averaged value of the raster for each latitude zone (using the default value of the argument dirXY) with time on the vertical axis.

```
hovmoller(SISdm)
```

On the other hand, this 2D matrix can be conceived as a multivariate time series with each aggregated zone conforming to a different variable of the time series. This approach is followed by the xyplot (Figure 19.7) and horizonplot (Figure 19.8) methods, which reproduce the procedures described in Chapter 3 to display multivariate time series.

```
xyplot(SISdm, auto.key = list(space = 'right'))
```

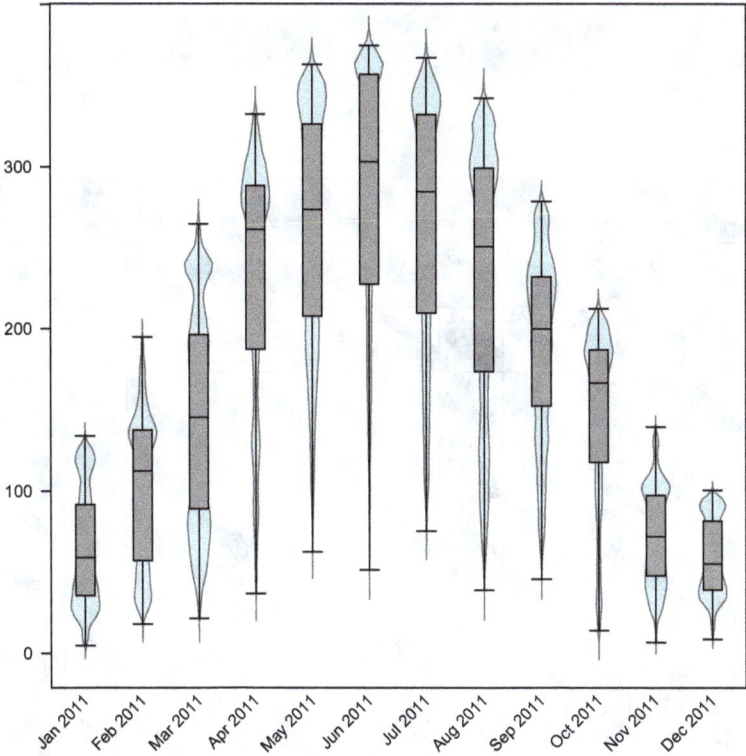

FIGURE 19.4: Violin plot of monthly distribution of solar radiation.

```
horizonplot(SISdm, digits = 1,
            col.regions = rev(brewer.pal(n = 6, 'PuOr')),
            xlab = '', ylab = 'Latitude')
```

These three figures highlight the seasonal behavior of the solar radiation, with higher values during the summer. It is interesting to note that (Figure 19.8) the radiation values around the equinoxes fluctuate near the yearly average value of each latitude region.

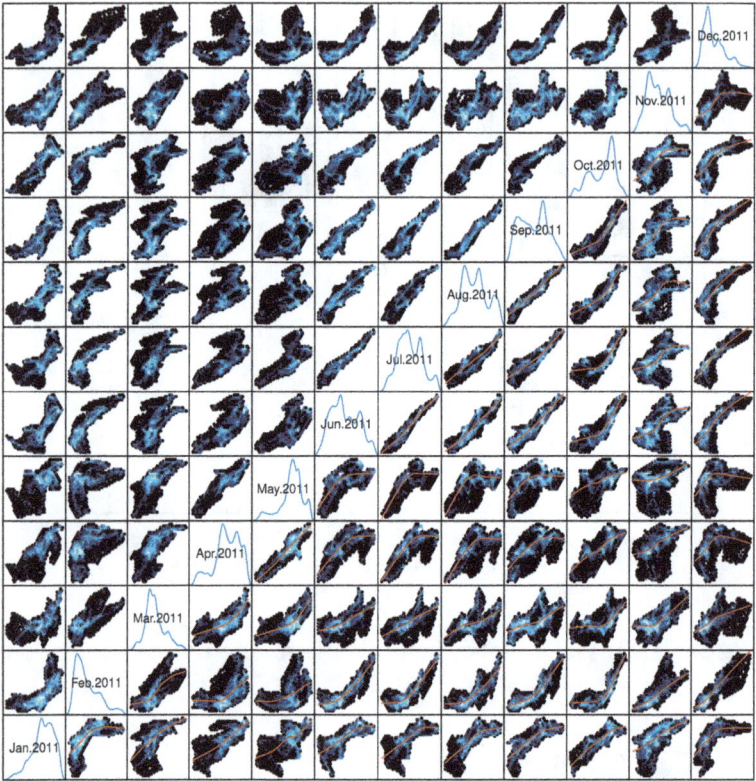

FIGURE 19.5: Scatterplot matrix of monthly averages together with their kernel density estimations in the diagonal frames.

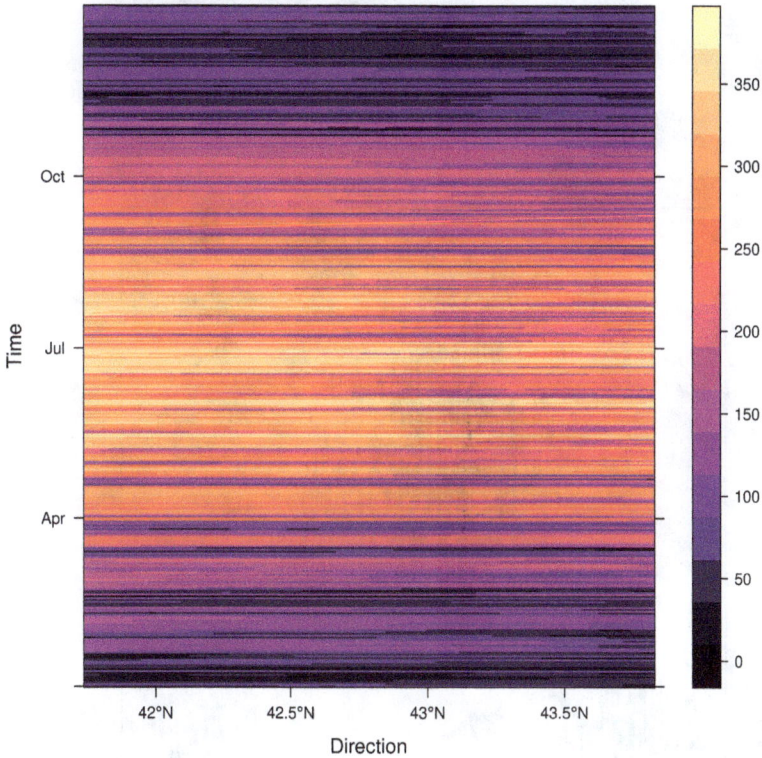

FIGURE 19.6: Hovmöller graphic displaying the time evolution of the average solar radiation for each latitude zone.

19.4.1 Interactive graphics: cubeView

Figure 19.6 reduces the 3D raster to a 2D matrix with spatial aggregation following a certain direction. The cubeview package[1] publishes a function with the same name to represent this 3D raster without prior aggregation as an interactive cube.

This cube can be freely rotated so that different Hövmoller views are possible. Visible layers can be selected using arrow keys (left-right for the x-axis, up-down for the y-axis), and PageUp-PageDown keys for the z-axis. Using the mouse, the cube can be rotated with the left button, moved

[1]This package only works with the raster package.

257

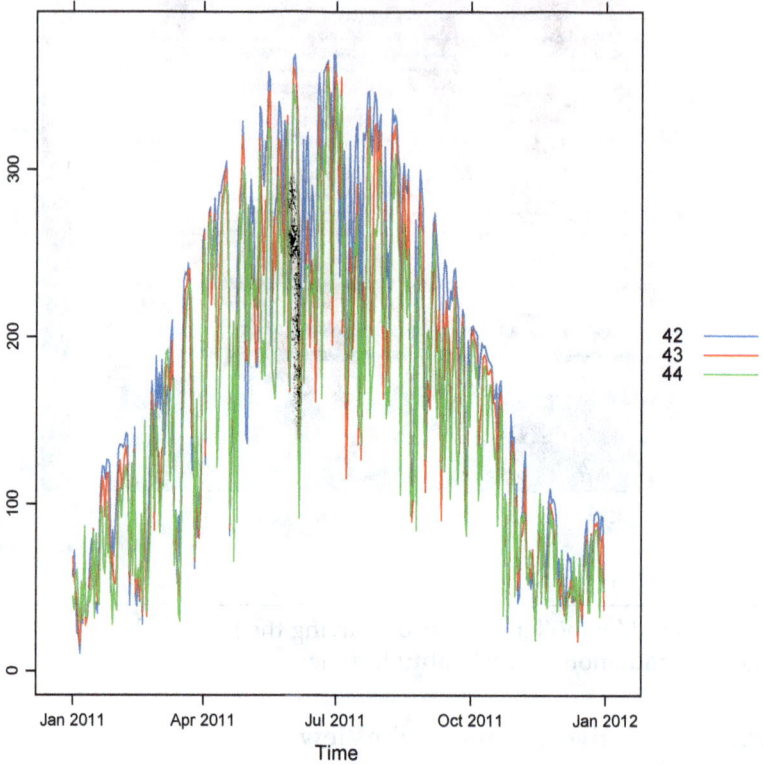

FIGURE 19.7: Time graph of the average solar radiation for each latitude zone. Each line represents a latitude band.

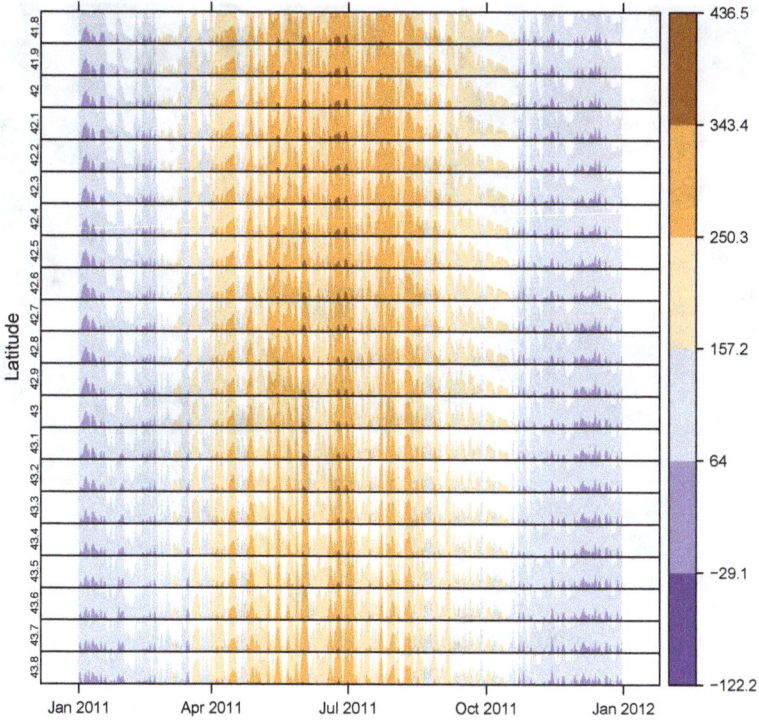

FIGURE 19.8: Horizon graph of the average solar radiation for each latitude zone.

with the right button, and zoomed using the mouse wheel. Figure 19.9 shows a snapshot of the cube produced with the next code.

```
library("cubeview")

## cubeview has problems if the Raster*
## is not stored in memory
SISdm <- readAll(SISdm)

cubeview(SISdm)
```

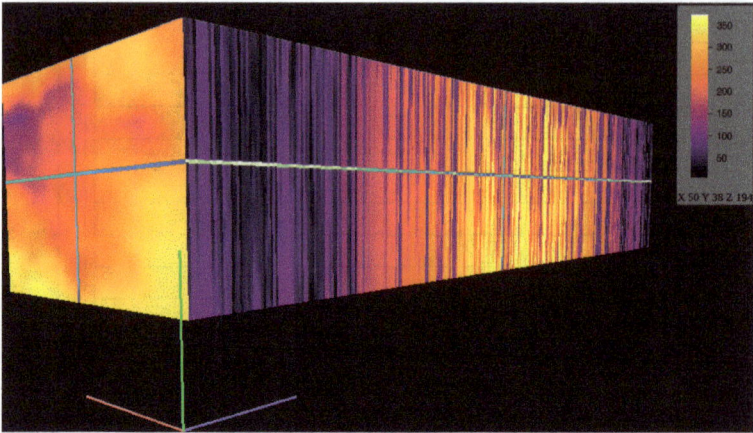

FIGURE 19.9: Snapshot of an interactive cube displaying a 3D raster.

19.5   Animation

This section uses animation to visualize changes of a raster variable over time. The procedure is quite simple:

- Plot each layer of the raster to produce a collection of graphic files.

- Join these files as a sequence of frames with a suitable tool (for example, ffmpeg[2]) to create a movie file[3, 4].

The effectiveness of this visualization procedure is partly related to the similitude between consecutive frames. If the frames of the sequence diverge excessively from one to another, the user will experience difficulties to perceive any relationship between them. On the other hand, if the transitions between layers are smooth enough, the frames will be perceived as conforming to a whole story; and, moreover, the user will be able to spot both the stable patterns and the important variations.

---

[2]http://www.ffmpeg.org/

[3]The animation package (Xie 2013) defines several functions to wrap ffmpeg and convert from ImageMagick.

[4]An alternative method is the LaTeX animate package, which provides an interface to create portable JavaScript-driven PDF animations from rasterized image files.

19.5.1 Data

The daily solar radiation CM-SAF data do not meet the condition of a smooth transition between layers. The changes between the consecutive snapshots of daily radiation are too abrupt to be glued one after another. We will work with a different dataset in this section.

The THREDSS server[5] of Meteogalicia[6] provides access through different protocols to the output of a Weather Research and Forecasting (WRF) model, a mesoscale numerical weather prediction system. Among the set of available variables, we will use the forecast of hourly cloud cover at low and mid levels. This space-time raster has a time horizon of 96 hours and a spatial resolution of 12 kilometers.

```
library("raster")
library("rasterVis")

cft <- brick("data/SpatioTime/cft_20130417_0000.nc")
## set projection
projLCC2d <- "+proj=lcc +lon_0=-14.1 +lat_0=34.823 +lat_1=43 +lat
    _2=43 +x_0=536402.3 +y_0=-18558.61 +units=km +ellps=WGS84"
projection(cft) <- projLCC2d
##set time index
timeIndex <- seq(as.POSIXct("2013-04-17 01:00:00", tz = "UTC"),
    length = 96, by = "hour")
cft <- setZ(cft, timeIndex)
names(cft) <- format(timeIndex, "D%d_H%H")
```

19.5.2 Spatial Context: Administrative Boundaries

Let's provide the spatial context with the countries boundaries, obtained from the Natural Earth database with the rnaturalearth package. The result of the query is a sf object. After being cropped to the extent of the raster and transformed to its projection, it is converted to a Spatial* object to be displayed with sp.lines.

```
library("rnaturalearth")
library("sf")
library("sp")

world <- ne_countries(scale = "medium")
```

---

[5]http://mandeo.meteogalicia.es/thredds/catalogos/WRF_2D/catalog.html
[6]http://www.meteogalicia.es

```
## Project the extent of the cft raster to longitude-latitude,
    because
## rnaturalearth works with it.
cftLL <- projectExtent(cft, crs(world))
## Crop...
boundaries <- st_crop(world, cftLL)
## ... and project to the projection of the cft object
boundaries <- st_transform(boundaries, crs(cft))
## Finally, convert to a Spatial* object
boundaries <- as(boundaries, "Spatial")
```

19.5.3 Producing the Frames and the Movie

The next step is to produce the collection of frames. We will create a file with each layer of the RasterBrick using the levelplot function. This function provides the argument layout to control the arrangement of a multipanel display. If it is set to c(1,1), a different page is created for each layer.

```
library("RColorBrewer")
library("latticeExtra")

cloudTheme <- rasterTheme(region = brewer.pal(n = 9, 'Blues'))

tmp <- tempdir()
trellis.device(png, file = paste0(tmp, "/Rplot%02d.png"),
               res = 300, width = 1500, height = 1500)
levelplot(cft, layout = c(1, 1),
          par.settings = cloudTheme,
          scales=list(draw=FALSE)) +
   layer(sp.lines(boundaries, lwd = 0.6))
dev.off()
```

A suitable tool to concatenate these frames and create the movie is ffmpeg, a free cross-platform software to record, convert, and stream audio and video[2]. The resulting movie is available from the book website.

```
old <- setwd(tmp)
## Create a movie with ffmpeg ...
system2("ffmpeg",
        c("-r 6", ## with 6 frames per second
          "-i Rplot%02d.png", ## using the previous files
          "-b:v 300k", ## with a bitrate of 300kbs
          "output.mp4")
```

```
        )
file.remove(dir(pattern = "Rplot"))
file.copy("output.mp4", paste0(old, "/figs/SpatioTime/cft.mp4"),
    overwrite = TRUE)
setwd(old)
```

19.5.4 Static Image

Figure 19.10 shows a sequence of twenty-four snapshots (second day of the forecast series) of the movie. This graphic is also created with `levelplot` but now using the argument `layers` to choose a subset of the layers and with a different value for `layout` to display a matrix of twenty-four panels.

```
levelplot(cft,
        layers = 25:48, ## Layers to display (second day)
        layout = c(6, 4), ## Layout of 6 columns and 4 rows
        par.settings = cloudTheme,
        scales=list(draw=FALSE),
        names.attr = paste0(sprintf("%02d", 1:24), "h"),
        panel = panel.levelplot.raster) +
    layer(sp.lines(boundaries, lwd = 0.6))
```

The movie and the static image are complementary tools and should be used together. Watching the movie, you will perceive the cloud transit from Galicia to the Pyrenees gradually dissolving over the Cantabrian region. On the other hand, with Figure 19.10 you can locate the position of a group of clouds in a certain hour and simultaneously observe the relationship of that position with the evolution during that period. With the movie, you will concentrate your attention on the movement. With small multiple pictures, your focus will be on positions and relations. You should use both graphical tools to grasp the entire 3D dataset.

19.5.5 3D animation

In section 11.4.1, an interactive 3D plot of a Digital Elevation Model was produced with the `rgl` package, a visualization device system for R using OpenGL as the rendering backend. With the next code, this package generates a 3D animation depicting the cloud evolution over time.

```
library("rgl")
```

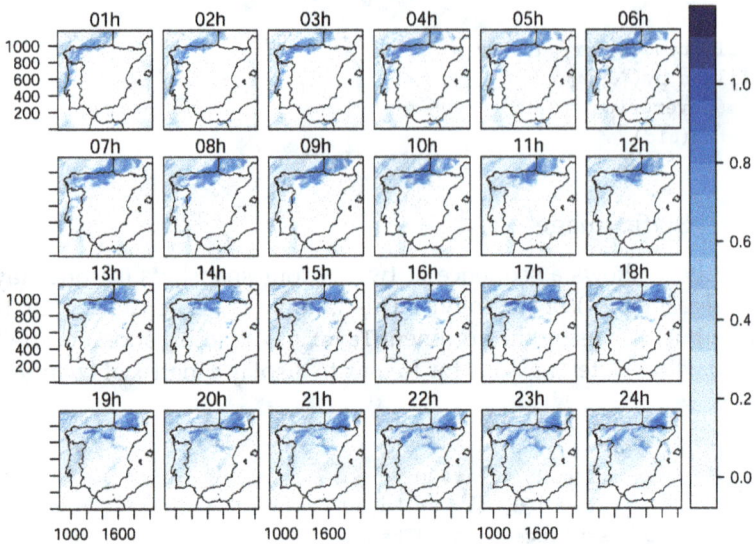

FIGURE 19.10: Forecast of hourly cloud cover at low and mid levels.

```
clear3d()

pal <- colorRampPalette(brewer.pal(n = 9, "Blues"))

N <- nlayers(cft)

ids <- lapply(seq_len(N),
         FUN = function(i)
             plot3D(cft[[i]],
                 maxpixels = 1e3,
                 col = pal,
                 adjust = FALSE, ## Disable automatic scaling
                     of xy axes.
                 zfac = 200)) ## Common z scale for all
                     graphics

library("manipulateWidget")

rglwidget() %>%
  playwidget(start = 0, stop = N,
         subsetControl(1, subsets = ids))
```

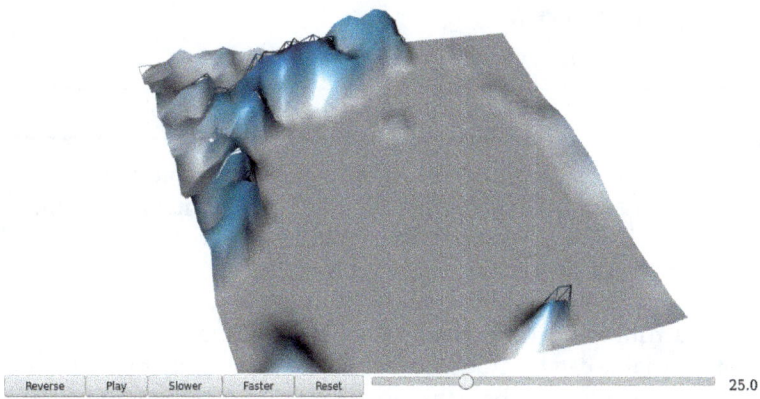

FIGURE 19.11: 3D animation of the forecast of hourly cloud cover at low and mid levels.

Chapter 20

Animation

Previous chapters include a section devoted to animations of point, raster, and aerial data. Notwithstanding, the animation technique is not restricted to displaying the evolution of a variable over time. Animations can also be constructed from the evolution of the position of a moving object (Section 20.1), from the sequence of changes in an attribute of the variable, for example, the user's view of a static variable can be gradually moved to create a fly-by animation (Section 20.2), and with the hill shading technique changing the sun position over time (Section 20.3).

The most relevant packages in this chapter are: sf for reading spatial vector data; raster for reading raster data; rnaturalearth, osmdata and move2 for retrieving data; rgl and rayshader for 3D visualization; and gganimate for animation with ggplot2.

DOI: 10.1201/9781003485384-20

20.1 Time Trajectory

A time trajectory is the evolution of the position of a moving object. It involves a series of positions usually captured by tracking devices such as GPS beacons. It can be represented with a static graphic, but the animation is a better solution to show the movement of the object.

The data of this section is extracted from the MoveBank platform[1], an online database of animal tracking data. It comes from a study funded by the "MERCURIO" project, which analyzes the trajectories of 10 lesser kestrels (small falcons) traveling from Senegal to Spain[2].

The move2 package[3] eases the import and handling of the data from MoveBank with R and extends the sf class with additional functionality, such as a time column. We will add spatial context with the administrative boundaries retrieved with the rnaturalearth package. Finally, the ggplot2 package will produce the static images with the geom_sf function, and the gganimate package[4] will create the animation.

```r
library("sf")
library("move2")
library("units")

library("rnaturalearth")

library("ggplot2")
library("gganimate")

## Movebank data
birds0 <- movebank_download_study(2398637362,
                     "license-md5"="74263192947
                              ce529c335a0ae72d7ead7")

sf_use_s2(FALSE) ## Needed for st_crop to work
## Natural Earth boundaries
boundaries <- ne_countries(scale = "large")
boundaries <- st_crop(boundaries, birds0)
```

---

[1] More information at: https://www.movebank.org/. You need an account (free) to download data: https://www.movebank.org/cms/movebank-registration

[2] https://www.movebank.org/cms/webapp?gwt_fragment=page=studies,path=study2398637362

[3] More information at https://bartk.gitlab.io/move2/. The package needs your username and password information to access the MoveBank data: https://bartk.gitlab.io/move2/articles/movebank.html

[4] https://gganimate.com/

```
## Filter the data: speed higher than 2 m/s; remove year 2022
    data.
birds <- subset(birds0,
                ground_speed > set_units(2L, "m/s") &
                timestamp >= as.POSIXct("2023-01-01"))

## Add a column with month values
birds$month <- as.numeric(format(mt_time(birds), "%m"))
```

The first attempt to display the trajectories is shown in Figure 20.1. Each individual is represented by a different color, and panels divide the data by months.

```
ggplot() +
    geom_sf(data = boundaries) +
    geom_sf(data = birds,
            aes(color = individual_local_identifier),
            alpha = 0.1) +
    guides(colour = guide_legend(override.aes = list(alpha = 1)))
    +
    theme_linedraw() +
    facet_wrap(~ month, nrow = 2)
```

The dataset contains the positions of the birds along with additional variables (speed, direction) captured by the tracking devices. As shown in Figure 20.2, these variables provide valuable insights into the trajectory data. This figure is a polar representation of the histogram of flight directions, using a different color for each speed class.

```
birds$speed <- cut(birds$ground_speed,
                   breaks = c(2, 5, 10, 15, 35))
ggplot() +
    coord_polar(start = 0) +
    geom_histogram(data = birds,
                   aes(x = set_units(heading, "degrees"),
                       fill = speed),
                   breaks = set_units(seq(0, 360, by = 10L), "degrees
                       "),
                   position = position_stack(reverse = TRUE)) +
    scale_x_units(name = NULL,
                  limits = set_units(c(0L, 360), "degrees"),
                  breaks = (0:4) * 90L) +
    ylab("") +
    facet_wrap(~ month, nrow = 2) +
    scale_fill_ordinal("Speed") +
```

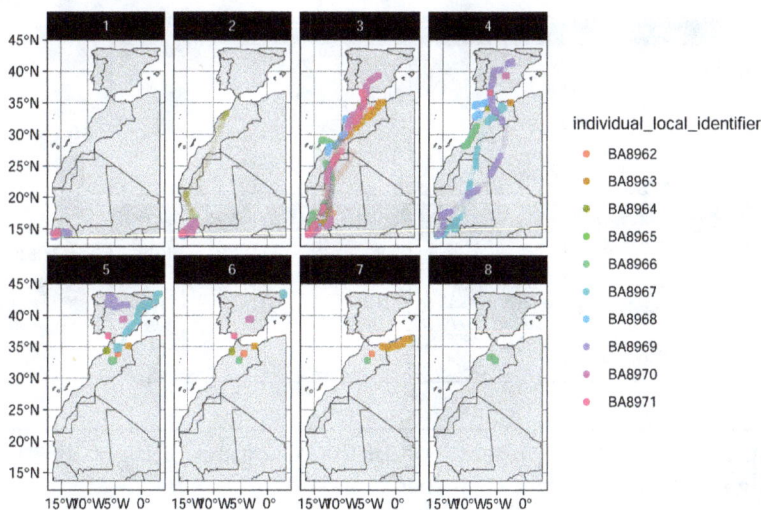

FIGURE 20.1: Trajectories of the lesser kestrels in the different months included in the study. Each individual is represented by a different color.

```
theme_linedraw()
```

Based on these previous figures, the animation will be focused on the trajectories registered during March. Two functions of the gganimate package must be noted here:

- transition_time splits the data into multiple states according to the values of the timestamp column and then generates a sequence inserting frames between the states for continuity.

- shadow_wake displays a wake after each position, showing previous states of the trajectory. Its numeric argument is the length of the wake as a percentage of the total length of the animation.

The animation is available at the book webpage. Figure 20.3 displays one of its frames.

```
birdsMarch <- subset(birds,
                month == 3)
p <- ggplot() +
    geom_sf(data = boundaries) +
    geom_sf(data = birdsMarch,
```

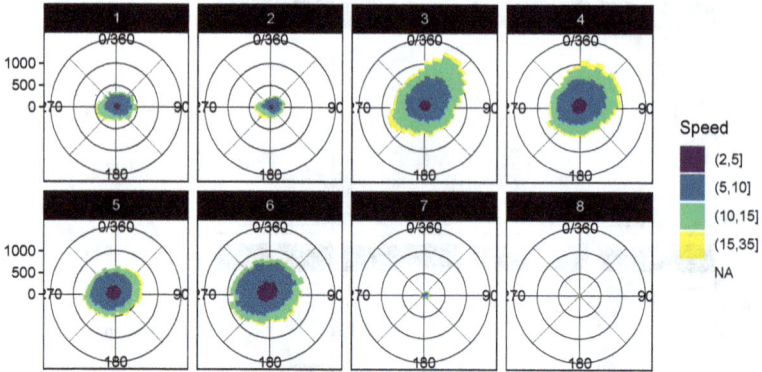

FIGURE 20.2: Polar representation of the histogram of flight directions, using a different color for each speed class.

```
            aes(colour = individual_local_identifier),
            size = 3) +
    theme_bw() +
    xlab("Longitude") + ylab("Latitude") +
    ggtitle("{format(frame_time, format = '%Y-%m-%d %H:%M:%S')}")
        +
    transition_time(timestamp) +
    shadow_wake(0.8)

animate(p, nframes = 300)
```

20.2 Fly-by Animation

In this section the `rgl` package is used to generate a fly-by animation over the Earth as an example of an animation depicting changes of a spatial attribute.

20.2.0.1 Basic 3D Earth

Firstly, a basic 3D Earth at night is created with the `surface3d` function, using the night lights images published by NASA[5].

---

[5]The page "Out of the Blue and Into the Black: New Views of the Earth at Night", https://earthobservatory.nasa.gov/Features/IntotheBlack/, provides detailed information about the Earth at Night maps.

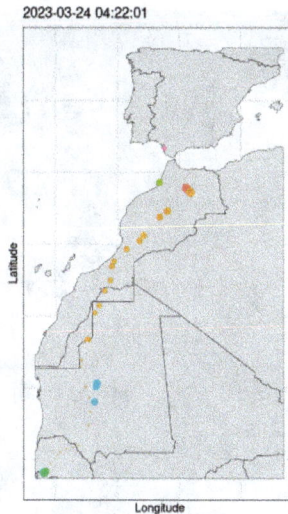

FIGURE 20.3: Snapshot of the animation of the trajectories.

```r
library("rgl")
library("magick") ## needed to import the texture

## Opens the OpenGL device with a black background
open3d()
bg3d("black")

## XYZ coordinates of a sphere
lat <- seq(-90, 90, len = 100) * pi/180
long <- seq(-180, 180, len = 100) * pi/180
r <- 6378.1 # radius of Earth in km
x <- outer(long, lat, FUN = function(x, y) r * cos(y) * cos(x))
y <- outer(long, lat, FUN = function(x, y) r * cos(y) * sin(x))
z <- outer(long, lat, FUN = function(x, y) r * sin(y))

## Read, scale, and convert the image
nightLightsJPG <- image_read("https://eoimages.gsfc.nasa.gov/
    images/imagerecords/79000/79765/dnb_land_ocean_
    ice.2012.13500x6750.jpg")
nightLightsJPG <- image_scale(nightLightsJPG, "8192") ##
    surface3d reads files up to 8192x8192
nightLights <- image_write(nightLightsJPG, tempfile(),
```

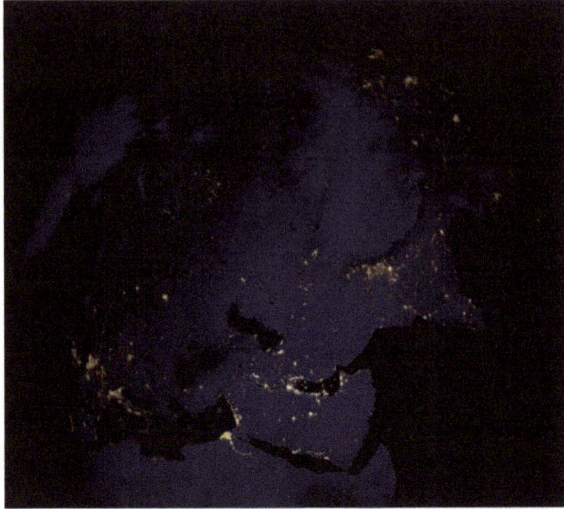

FIGURE 20.4: Snapshot of the WebGL figure created with `rglwidget`.

```
                                format = "png") ## Only the png format is
                                supported
## Display the sphere with the image superimposed
surface3d(-x, -z, y,
         texture = nightLights,
         specular = "black", col = "white")
```

This OpenGL object can be exported to different formats. For example, Figure 20.4 shows a snapshot of the WebGL figure created with `rglwidget`:

```
rglwidget()
```

20.2.0.2 Define the Locations

Once the Earth is represented with the sphere and the superimposed image, the fly-by animation is defined with a set of locations to be visited:

```
cities <- rbind(c("Madrid", "Spain"),
               c("Tokyo", "Japan"),
               c("Sidney", "Australia"),
               c("Sao Paulo", "Brazil"),
               c("New York", "USA"))
```

```r
cities <- as.data.frame(cities)
names(cities) <- c("city", "country")
```

The latitude and longitude coordinates of these cities can be obtained through the Nominatim service of OpenStreetMap with the package os-mdata. The function getbb of this package returns the bounding box of a city. The auxiliary function geocode defined below computes the center of each bounding box.

```r
library("osmdata")

geocode <- function(x) {
  place <- paste(x, collapse = ", ")
  bb <- getbb(place)
  center <- c(sum(bb[1, ])/2, sum(bb[2, ]/2))
  center
}

points <- apply(cities, 1, geocode)
points <- t(points)
colnames(points) <- c("lon", "lat")

cities <- cbind(cities, points)
```

20.2.0.3   Generate the Route

The next step computes the intermediate points between each pair of locations. The geosphere package provides the gcIntermediate function for this task:

```r
library("geosphere")

## When arriving or departing include a progressive zoom with 100
## frames
zoomIn <- seq(.3, .1, length = 100)
zoomOut <- seq(.1, .3, length = 100)

## First point of the route
route <- data.frame(lon = cities[1, "lon"],
                    lat = points[1, "lat"],
                    zoom = zoomIn,
                    name = cities[1, "city"],
                    action = "arrive")
```

```r
## This loop visits each location included in the "points" set
## generating the route.
for (i in 1:(nrow(cities) - 1)) {

  p1 <- cities[i,]
  p2 <- cities[i + 1,]
  ## Initial location
  departure <- data.frame(lon = p1$lon,
                          lat = p1$lat,
                          zoom = zoomOut,
                          name = p1$city,
                          action = "depart")


  ## Travel between two points: Compute 100 points between the
  ## initial and the final locations.
  routePart <- gcIntermediate(p1[, c("lon", "lat")],
                              p2[, c("lon", "lat")],
                              n = 100)
  routePart <- data.frame(routePart)
  routePart$zoom <- 0.3
  routePart$name <- ""
  routePart$action <- "travel"

  ## Final location
  arrival <- data.frame(lon = p2$lon,
                        lat = p2$lat,
                        zoom = zoomIn,
                        name = p2$city,
                        action = "arrive")
  ## Complete route: initial, intermediate, and final locations.
  routePart <- rbind(departure, routePart, arrival)
  route <- rbind(route, routePart)
}

## Close the travel
route <- rbind(route,
               data.frame(lon = cities[i + 1, "lon"],
                          lat = cities[i + 1, "lat"],
                          zoom = zoomOut,
                          name = cities[i+1, "city"],
                          action = "depart"))

summary(route)
```

```
      lon               lat                zoom              name
Min.   :-178.490   Min.   :-74.728   Min.   :0.1000   Length:1500
1st Qu.: -62.896   1st Qu.:-23.683   1st Qu.:0.1687   Class :character
Median :  -3.703   Median : 28.056   Median :0.2374   Mode  :character
Mean   :  28.233   Mean   :  9.974   Mean   :0.2267
3rd Qu.: 145.030   3rd Qu.: 40.697   3rd Qu.:0.3000
Max.   : 179.446   Max.   : 68.796   Max.   :0.3000
    action
Length:1500
Class :character
Mode  :character
```

20.2.0.4 Produce the Frames

Finally, this matrix of points is used to change the viewpoint of the OpenGL scene with the `view3d` function. The `travel` function wraps this function to automate the process with the `movie3d` function and returns a list of parameters of the scene. Figure 20.5 shows an example of a frame produced with this function.

```
## Function to move the viewpoint in the RGL scene according to
   the
## information included in the route (position and zoom).
travel <- function(tt) {
  point <- route[tt,]
  view3d(theta = -90 + point$lon,
         phi = point$lat,
         zoom = point$zoom)
  par3d()
}
```

```
## Example of usage of travel
## Frame no.1200
travel(1200)
rgl.snapshot("figs/SpatioTime/rgl_travel1200.png")
```

The `movie3d` accepts a function returning a list of parameters, `travel` in our code, to modify the RGL scene. It creates a snapshot at each step and paste these snapshots as frames of a movie.

```
movie3d(travel,
        duration = nrow(route),
        startTime = 1, fps = 1,
```

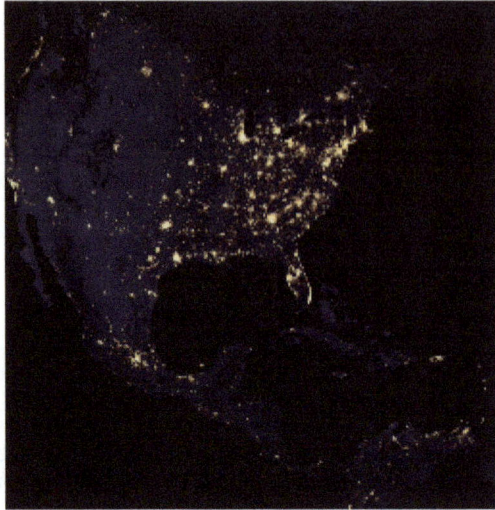

FIGURE 20.5: Example of usage of the `travel` function (frame no. 1200).

```
type = "mp4", clean = FALSE,
webshot = FALSE)
```

20.3   Hill shading animation

The hill shading technique has been previously explained in sections 11.1.1 and 14.3. In those examples, the sun's position was fixed. We will use it again, changing the sun position across the day and tracing the shades with the `rayshader` package[6]. The sun coordinates will be computed with the `suntools` package, and the `gif` animation will be produced with the `gifski` package.

```
library("rayshader")
library("parallel")
library("suntools")
library("raster")
library("gifski")
```

---

[6]https://www.rayshader.com/

For this section we will work with the same Digital Elevation Model of section 14.3. It is read with `raster` and converted to a matrix to be used with `rayshader`.

```
demCedeira <- raster('data/Spatial/demCedeira')
DEM <- raster_to_matrix(demCedeira)
```

The function `detect_water` of this package can detect areas of water in a DEM.

```
water <- detect_water(DEM)
```

The sun position is computed with the `solarpos` function of the `suntools` package. This function requires a location, which is the center of the DEM, and a timestamp. The position is computed from the sunrise to the sunset, approximately.

```
lonlat <- matrix(c((xmax(demCedeira) + xmin(demCedeira))/2,
                   (ymax(demCedeira) + ymin(demCedeira))/2),
               nrow = 1)

tt <- seq(as.POSIXct("2024-06-01 07:00:00", tz = "Europe/Madrid")
    ,
          as.POSIXct("2024-06-01 21:00:00", tz = "Europe/Madrid"),
          by = "15 min")
sun <- lapply(tt, function(x) solarpos(lonlat, x))
```

Everything is ready for the hill shading computation. Three functions are to be noted: `sphere_shade` maps a texture to a hillshade by spherical mapping; `ray_shade` uses the specified light directions to calculate a global shadow map for the elevation matrix; and `add_shadow` combines these shadow maps. The `mclapply` function of the `parallel` package is used here for speeding up the loop with parallel computing.

```
ncores <- detectCores()

hillshades <- mclapply(sun, function(ang)
{
    DEM %>%
        sphere_shade(texture = "imhof1", sunangle = ang[1]) %>%
        add_water(water, color = "imhof1") %>%
        add_shadow(ray_shade(DEM,
                        sunangle = ang[1],
                        sunaltitude = ang[2]),
                   0.75)
}, mc.cores = ncores)
```

The `hillshades` object is a list of matrices, each containing a shade map for a certain sun position. The `plot_3d` function can display this map over the DEM map, and the `render_snapshot` function captures this 3D view to an image file. By iterating through the list and running these functions in a loop, you can generate a series of images that the `gifski` function can compile into a GIF animation. Figure 20.6 shows a snapshot of the animation, which is available at the book webpage.

```
old <- setwd(tempdir())

idx <- seq_along(hillshades)

for(i in idx)
{
    plot_3d(heightmap = DEM,
            hillshade = hillshades[[i]],
            zscale = 5,
            fov = 45,
            theta = 0,
            zoom = 0.75,
            phi = 45,
            windowsize = c(1000, 800))
    render_snapshot(filename = paste0(i, ".png"),
                    title_text = tt[i])
}

gifski(png_files = paste0(idx, ".png"),
       gif_file = "cedeira.gif",
       delay = 1/5)

setwd(old)
```

2024-06-01 12:30:00

FIGURE 20.6: Snapshot of the hill shade animation.

Part IV

Glossary, Bibliography, and Index

Glossary

CM-SAF: Satellite Application Facility on Climate Monitoring.

CRAN: Comprehensive R Archive Network.

DEM: Digital Elevation Model.

Geo-TIFF: A public domain metadata standard that allows georeferencing information to be embedded within a TIFF file.

GeoJSON: A format for encoding a variety of geographic data structures.

GIS: Geographic Information Systems.

IDW: Inverse Distance Weighted interpolation.

KML: Keyhole Markup Language, an XML notation for expressing geographic annotation and visualization within Internet-based, two-dimensional maps and three-dimensional Earth browsers.

NEO-NASA: NASA Earth Observations, part of NASA's Earth Observing System (EOS).

NetCDF: Network Common Data Form, a set of software libraries and self-describing, machine-independent data formats that support the creation, access, and sharing of array-oriented scientific data.

RasterBrick: A class to represent multilayer (variable) raster data with the raster package.

RasterLayer: A class to represent single-layer (variable) raster data with the raster package.

RasterStack: A class to represent multilayer (variable) raster data with the raster package.

shapefile: A geospatial vector data format developed by Esri that can store vector features (points, lines, and polygons) and attributes that describe these features (name of the location, temperature, etc.).

SIS: Shortwave incoming solar radiation.

SpatialLinesDataFrame: Class for spatial attributes consisting of sets of lines, where each set of lines relates to an attribute row in a data.frame.

SpatialPixelsDataFrame: Class for spatial attributes that have spatial locations on a regular grid.

SpatialPointsDataFrame: Class for spatial attributes that have spatial point locations.

SpatialPolygonsDataFrame: Class to hold polygons with attributes.

SpatRaster: A class to represent raster data with the terra package.

SpatVector: A class to represent spatial vector data with the terra package.

STL: File format that encodes the surface geometry of a 3D object using tessellation.

SVG: Scalable Vector Graphics.

TIFF: Tagged Image File Format, a computer file format for storing raster graphics images.

WRF: Weather Research and Forecasting model.

Bibliography

Antonanzas-Torres, F., F. Cañizares, and O. Perpiñán (2013). "Comparative assessment of global irradiation from a satellite estimate model (CM SAF) and on-ground measurements (SIAR): A Spanish case study". In: *Renewable and Sustainable Energy Reviews* 21.0, pp. 248–261. ISSN: 1364-0321. DOI: 10.1016/j.rser.2012.12.033. URL: https://github.com/oscarperpinan/CMSAF-SIAR.

Bivand, R. (2023). *classInt: Choose Univariate Class Intervals*. R package version 0.4-10. URL: http://CRAN.R-project.org/package=classInt.

Bivand, R., E. J. Pebesma, and V. Gomez-Rubio (2013). *Applied Spatial Data Analysis with R*. Springer, New York. URL: http://www.asdar-book.org/.

Byron, L. and M. Wattenberg (2008). *Stacked Graphs – Geometry & Aesthetics*. Tech. rep. URL: https://leebyron.com/streamgraph/stackedgraphs_byron_wattenberg.pdf.

Cairo, A. (2012). *The Functional Art: An Introduction to Information Graphics and Visualization*. New Riders Publishing, Aarhus, Denmark.

Carr, D., N. Lewin-Koh, and M. Maechler (2024). *hexbin: Hexagonal Binning Routines*. R package version 1.28.5. URL: https://CRAN.R-project.org/package=hexbin.

Carr, D. B. et al. (1987). "Scatterplot Matrix Techniques for Large N". English. In: *Journal of the American Statistical Association* 82.398, pp. 424–436. ISSN: 01621459. URL: http://www.jstor.org/stable/2289444.

Chambers, J. M. (2008). *Software for Data Analysis: Programming with R*. ISBN 978-0-387-75935-7. Springer, New York.

Chatfield, C. (2019). *The Analysis of Time Series: An Introduction*. Chapman & Hall/CRC Texts in Statistical Science. CRC Press. ISBN: 9781351259446.

Cleveland, W. S. (1993). *Visualizing Data*. Summit, NJ: Hobart Press.

— (1994). *The Elements of Graphing Data*. Murray Hill, NJ.: AT&T, Bell Laboratories.

Cleveland, W. S. and R. McGill (1984). "Graphical Perception: Theory, Experimentation, and Application to the Development of Graphical Methods". In: *Journal of the American Statistical Association* 79.387, pp. 531–554. ISSN: 01621459. URL: http://www.jstor.org/stable/2288400.

CM SAF (2024). *The Satellite Application Facility on Climate Monitoring*. http://www.cmsaf.eu.

Cressie, N. and C.K. Wikle (2015). *Statistics for Spatio-Temporal Data*. Wiley Series in Probability and Statistics. Wiley, New York. ISBN: 9781119243045.

Dent, B., J. Torguson, and T. Hodler (2008). *Cartography: Thematic Map Design*. McGraw-Hill Education, New York. ISBN: 9780072943825.

Few, S. (2007). *Visualizing Change: An Innovation in Time-Series Analysis*. Tech. rep. Perceptual Edge, Berkeley, CA. URL: http://www.perceptualedge.com/articles/visual_business_intelligence/visualizing_change.pdf.

— (2008). *Time on the Horizon*. Tech. rep. Perceptual Edge, Berkeley, CA. URL: http://www.perceptualedge.com/articles/visual_business_intelligence/time_on_the_horizon.pdf.

Friendly, M. and D. Denis (2005). "The early origins and development of the scatterplot". In: *Journal of the History of the Behavioral Sciences* 41.2, pp. 103–130. ISSN: 1520-6696. DOI: 10.1002/jhbs.20078.

Grothendieck, G. and T. Petzoldt (2004). "R Help Desk: Date and Time Classes in R". In: *R News* 4.1, pp. 29–32. URL: http://CRAN.R-project.org/doc/Rnews/Rnews_2004-1.pdf.

Harrower, M. and S. I. Fabrikant (2008). "The Role of Map Animation in Geographic Visualization". In: *Geographic Visualization: Concepts, Tools and Applications*. Ed. by M. Dodge, M. McDerby, and Turner M. Chichester, UK: Wiley, pp. 49–65. URL: http://www.zora.uzh.ch/8979/.

Havre, S. et al. (2002). "ThemeRiver: Visualizing Thematic Changes in Large Document Collections". In: *IEEE Transactions on Visualization and Computer Graphics* 8.1, pp. 9–20. ISSN: 1077-2626. DOI: 10.1109/2945.981848.

Heer, J. and M. Agrawala (2006). "Multi-Scale Banking to 45 Degrees". In: *IEEE Transactions on Visualization and Computer Graphics* 12.5, pp. 701–708. ISSN: 1077-2626. DOI: 10.1109/TVCG.2006.163. URL: http://vis.berkeley.edu/papers/banking/2006-Banking-InfoVis.pdf.

Heer, J., M. Bostock, and V. Ogievetsky (2010). "A tour through the visualization zoo". In: *Communications of the ACM* 53.6, pp. 59–67. ISSN: 0001-0782. DOI: 10.1145/1743546.1743567. URL: http://doi.acm.org/10.1145/1743546.1743567.

Heer, J., N. Kong, and M. Agrawala (2009). "Sizing the Horizon: The Effects of Chart Size and Layering on the Graphical Perception of Time Series Visualizations". In: *ACM Human Factors in Computing Systems (CHI)*, pp. 1303–1312. URL: http://vis.berkeley.edu/papers/horizon/2009-TimeSeries-CHI.pdf.

Hengl, T. (2009). *A Practical Guide to Geostatistical Mapping.* University of Amsterdam, Amsterdam. URL: https://publications.jrc.ec.europa.eu/repository/handle/JRC38153/.

Hijmans, R. J. (2024a). *raster: Geographic Data Analysis and Modeling.* R package version 3.6-30. URL: https://CRAN.R-project.org/package=raster.

— (2024b). *terra: Spatial Data Analysis.* R package version 1.8-5. URL: https://CRAN.R-project.org/package=terra.

Hocking, T. D. (2024). *directlabels: Direct Labels for Multicolor Plots.* R package version 2024.1.21. URL: http://CRAN.R-project.org/package=directlabels.

Hovmöller, E. (1949). "The Trough-and-Ridge diagram". In: *Tellus* 1.2, pp. 62–66. ISSN: 2153-3490. DOI: 10.1111/j.2153-3490.1949.tb01260.x.

Ihaka, R. et al. (2024). *colorspace: Color Space Manipulation.* R package version 2.1-1. URL: http://CRAN.R-project.org/package=colorspace.

Kropotkin, P. (1906). *The Conquest of Bread.* G. P. Putnam's Sons. URL: http://en.wikisource.org/wiki/The_Conquest_of_Bread.

Lucchesi, L. R. and C. K. Wikle (2017). "Visualizing uncertainty in areal data with bivariate choropleth maps, map pixelation and glyph rotation". In: *Stat* 6.1, pp. 292–302. DOI: 10.1002/sta4.150.

MAGRAMA (2024). *Sistema de Información Agroclimática del Regadío.* https://servicio.mapa.gob.es/websiar/.

Meihoefer, H. J. (1969). "The Utility of the Circle as an Effective Cartographic Symbol". In: *Cartographica: The International Journal for Geographic Information and Geovisualization* 6 (2), pp. 104–117. DOI: 10.3138/J04Q-1K34-26X1-7244.

Mumford, L. (1934). *Technics and Civilization*. San Diego, CA.: Harcourt, Brace & Company, Inc.

Murrell, P. (2021). *R Graphics*. 3rd. The R Series. Boca Raton, FL.: Chapman & Hall/CRC.

Murrell, P. and S. Potter (2023). *gridSVG: Export grid graphics as SVG*. R package version 1.7-5. URL: http://CRAN.R-project.org/package=gridSVG.

Neuwirth, E. (2022). *RColorBrewer: ColorBrewer Palettes*. R package version 1.1-3. URL: http://CRAN.R-project.org/package=RColorBrewer.

Pebesma, E. (2012). "`spacetime`: Spatio-Temporal Data in R". In: *Journal of Statistical Software* 51.7, pp. 1–30. ISSN: 1548-7660. URL: http://www.jstatsoft.org/v51/i07.

— (2024). *sf: Simple Features for R*. R package version 1.0-19. URL: https://CRAN.R-project.org/package=sf.

Pebesma, E. J. (2004). "Multivariable Geostatistics in S: The gstat Package". In: *Computers and Geosciences* 30, pp. 683–691.

Pebesma, E. J. and R. Bivand (2005). "Classes and methods for spatial data in R". In: *R News* 5.2, pp. 9–13. URL: http://CRAN.R-project.org/doc/Rnews/.

Perpiñán, O. (2012). "solaR: Solar Radiation and Photovoltaic Systems with R". In: *Journal of Statistical Software* 50.9, pp. 1–32. URL: http://www.jstatsoft.org/v50/i09/.

Perpiñán, O. and R. Hijmans (2023). *rasterVis: Visualization Methods for the raster Package*. R package version 0.51.6. URL: https://oscarperpinan.codeberg.page/rastervis/.

Perpiñán, Oscar and Marcelo Pinho Almeida (2023). *meteoForecast*. R package version 0.56. URL: https://codeberg.org/oscarperpinan/meteoForecast.

Pinho-Almeida, M., O. Perpiñán, and L Narvarte (2015). "PV Power Forecast Using a Nonparametric PV Model". In: *Solar Energy* 115, pp. 354–368. ISSN: 0038092X. DOI: 10.1016/j.solener.2015.03.006. URL: https://oa.upm.es/34853/.

Posselt, R., R.W. Mueller, et al. (2012). "Remote sensing of solar surface radiation for climate monitoring—the CM SAF retrieval in international comparison". In: *Remote Sensing of Environment* 118.0, pp. 186–198. ISSN: 0034-4257. DOI: 10.1016/j.rse.2011.11.016.

Posselt, R., R. Müller, et al. (2011). *CM SAF Surface Radiation MVIRI Data Set 1.0 - Monthly Means / Daily Means / Hourly Means*. DOI: 10.5676/EUM_SAF_CM/RAD_MVIRI/V001. URL: http://dx.doi.org/10.5676/EUM_SAF_CM/RAD_MVIRI/V001.

R Core Team (2024). *R: A Language and Environment for Statistical Computing*. R Foundation for Statistical Computing. Vienna, Austria. URL: https://www.R-project.org/.

Ripley, B. D. and K. Hornik (2001). "Date-Time Classes". In: *R News* 1.2, pp. 8–11. URL: http://CRAN.R-project.org/doc/Rnews/Rnews_2001-2.pdf.

Rossini, A. J. et al. (2004). "Emacs Speaks Statistics: A Multiplatform, Multipackage Development Environment for Statistical Analysis". In: *Journal of Computational and Graphical Statistics* 13.1, pp. 247–261. DOI: 10.1198/1061860042985. eprint: http://www.tandfonline.com/doi/pdf/10.1198/1061860042985. URL: http://ess.r-project.org/.

Ryan, J. A. and J. M. Ulrich (2024). *xts: eXtensible Time Series*. R package version 0.14.1. URL: http://CRAN.R-project.org/package=xts.

Sarkar, D. (2008). *Lattice: Multivariate Data Visualization with R*. ISBN 978-0-387-75968-5. New York: Springer. URL: http://lmdvr.r-forge.r-project.org.

Sarkar, D. and F. Andrews (2022). *latticeExtra: Extra Graphical Utilities Based on lattice*. R package version 0.6-30. URL: http://CRAN.R-project.org/package=latticeExtra.

Schulte, E. et al. (2012). "A Multi-Language Computing Environment for Literate Programming and Reproducible Research". In: *Journal of Statistical Software* 46.3, pp. 1–24. ISSN: 1548-7660. URL: http://www.jstatsoft.org/v46/i03.

Schulz, J. et al. (2009). "Operational Climate Monitoring from Space: The EUMETSAT Satellite Application Facility on Climate Monitoring". In: *Atmospheric Chemistry and Physics* 9, pp. 1687–1709. DOI: 10.5194/acp-9-1687-2009. URL: http://www.atmos-chem-phys.net/9/1687/2009/acp-9-1687-2009.pdf.

Slocum, T. A. (2022). *Thematic Cartography and Geographic Visualization*. CRC Press. ISBN: 9780367712709.

Teickner, Henning, Edzer Pebesma, and Benedikt Graeler (2024). *sftime: Classes and Methods for Simple Feature Objects that Have a Time Column*. R package version 0.3.0. URL: https://CRAN.R-project.org/package=sftime.

Tufte, E. R. (1990). *Envisioning information*. Cheshire, CT.: Graphic Press.

— (2001). *The Visual Display of Quantitative Information*. Cheshire, CT.: Graphic Press.

Vaidyanathan, R. et al. (2023). *htmlwidgets: HTML Widgets for R*. R package version 1.6.4. URL: https://CRAN.R-project.org/package=htmlwidgets.

Ware, C. (2021). *Visual Thinking for Design*. Morgan Kaufmann. ISBN: 9780128235676.

Wegenkittl, R. and E. Gröller (1997). "Fast Oriented Line Integral Convolution for Vector Field Visualization via the Internet". In: *Proceedings of the 8th Conference on Visualization'97*. IEEE Computer Society Press, pp. 309–316. URL: https://www.cg.tuwien.ac.at/research/vis-dyn-syst/frolic/frolic_crc.pdf.

Wickham, H. (2016). *ggplot2: Elegant Graphics for Data Analysis*. Springer. URL: https://github.com/hadley/ggplot2-book/.

Wilkinson, L. (2005). *The Grammar of Graphics*. Springer.

Wills, G. (2012). *Visualizing Time: Designing Graphical Representations for Statistical Data*. Statistics and Computing. New York: Springer. ISBN: 9780387779065.

Xie, Y. (2013). "animation: An R Package for Creating Animations and Demonstrating Statistical Methods". In: *Journal of Statistical Software* 53.1, pp. 1–27. ISSN: 1548-7660. URL: http://www.jstatsoft.org/v53/i01.

Zeileis, A. and G. Grothendieck (2005). "zoo: S3 Infrastructure for Regular and Irregular Time Series". In: *Journal of Statistical Software* 14.6, pp. 1–27. URL: http://www.jstatsoft.org/v14/i06/.

Index

For Product Safety Concerns and Information please contact our EU
representative GPSR@taylorandfrancis.com
Taylor & Francis Verlag GmbH, Kaufingerstraße 24, 80331 München, Germany

www.ingramcontent.com/pod-product-compliance
Lightning Source LLC
Chambersburg PA
CBHW060447240326
41598CB00088B/3901